ATMOSPHERIC FRONTAL DYNAMICS

Atmospheric Frontal Dynamics is the first book to present self-contained and comprehensive coverage of the ubiquitous surface and upper-level fronts that are boundaries separating two air masses of different densities. Atmospheric fronts are a product of interaction between planetary, synoptic, meso, and small scales of motions. They are uniquely important in weather forecasting since many meteorological phenomena are closely associated with them. The discussions of the multifaceted aspects of basic frontal dynamics are based on illustrative analyses of a hierarchy of semi-geostrophic frontal models performed in physical, as opposed to geostrophic, coordinates. Such analyses enable us to systematically and quantitatively delineate the physical nature of different types of fronts in a relatively straightforward manner. They are further supplemented by two illustrative analyses using a state-of-the-art model. "Learning by doing" is the guiding principle behind the book, making it an asset to both students and instructors in atmospheric science and meteorology.

MANKIN MAK is Professor Emeritus in the Department of Atmospheric Sciences at the University of Illinois, Urbana-Champaign, where he has taught and researched atmospheric dynamics for many years. He completed his undergraduate study in engineering physics at the University of Toronto, Canada, followed by graduate study in meteorology at the Massachusetts Institute of Technology, USA. He has published extensively in many international journals of atmospheric sciences and served as an editor of the *Journal of Atmospheric Sciences*. He is the author of *Atmospheric Dynamics* (2011, Cambridge University Press). His current research interest focuses on atmospheric frontal dynamics.

ATMOSPHERIC FRONTAL DYNAMICS

MANKIN MAK
University of Illinois, Urbana-Champaign

CAMBRIDGE
UNIVERSITY PRESS

CAMBRIDGE
UNIVERSITY PRESS

University Printing House, Cambridge CB2 8BS, United Kingdom

One Liberty Plaza, 20th Floor, New York, NY 10006, USA

477 Williamstown Road, Port Melbourne, VIC 3207, Australia

4843/24, 2nd Floor, Ansari Road, Daryaganj, Delhi – 110002, India

79 Anson Road, #06-04/06, Singapore 079906

Cambridge University Press is part of the University of Cambridge.

It furthers the University's mission by disseminating knowledge in the pursuit of education, learning, and research at the highest international levels of excellence.

www.cambridge.org
Information on this title: www.cambridge.org/9781107133204
DOI: 10.1017/9781316459584

First published 2018

A catalogue record for this publication is available from the British Library.

Library of Congress Cataloging-in-Publication Data

Names: Mak, Mankin, 1939– author.
Title: Atmospheric frontal dynamics / Mankin Mak, University of Illinois at Urbana-Champaign.

Description: Cambridge : Cambridge University Press, 2017. | Includes bibliographical references and index.

Identifiers: LCCN 2017035910 | ISBN 9781107133204 (hardback : alk. paper)

Subjects: LCSH: Fronts (Meteorology) | Air masses. | Atmospheric physics.

Classification: LCC QC880.4.F7 M36 2017 | DDC 551.55/12–dc23 LC record available at https://lccn.loc.gov/2017035910

ISBN 978-1-107-13320-4 Hardback

This book is dedicated to
Professor Norman A. Phillips,
my 94-year-old former mentor (the memorable past),
Pauline (the wonderful present), and
Kiran Anying and Deven Zhonghan, my grandchildren
(the promising future).

Contents

Preface

If a weather forecaster in mid-latitude were told that he could have only one piece of information about the atmosphere before making a short-term prediction, what would he ask for? Chances are that he would ask for the position and latest movement of the fronts in the region of his concern because large changes in weather conditions leading to damaging events generally occur in the neighborhood of surface fronts (SFs). Also, large local exchange of mass takes place between the stratosphere and troposphere at an upper-level front (ULF), which is an integral part of a frontal system. A frontal system has a complicated structure involving nonlinear interaction among the planetary, synoptic, meso-, and small-scale components of the atmospheric circulation. It is no wonder that fronts have been the subject of continuing research for over a century. One might have expected to find extensive coverage of the fundamental dynamics of fronts in the textbooks of atmospheric sciences. Surprisingly, a survey of the well-known textbooks on atmospheric dynamics suggests otherwise. Most of them do not address this topic at all. Some only have qualitative and kinematic discussions of fronts and frontogenesis. A handful do cover frontal dynamics, but only in the context of a canonical 2D semi-geostrophic (SG) model. This survey is elaborated on in Section 1.5.

Frontal dynamics are discussed in this book in conjunction with a number of self-contained illustrative analyses using progressively more complex frontal models. We first discuss why an SG model is the simplest framework that embodies the fundamental dynamical processes responsible for frontogenesis. We then show how to formulate and perform analyses with two 2D and one 3D generic SG models. The impacts of diabatic heating in SG models are addressed with a parameterized form of such heating. Finally, we use a state-of-the-art high-resolution model to simulate SF and ULF at the same time. The specific objectives are to delineate the dynamics of upper-level frontogenesis as well as the dynamics behind the distinct differences between cold and warm SF. It also enables us to examine whether, how, and why a ULF and an SF could synergistically interact. Enough details of each illustrative analysis are elaborated on that readers would be in a position to duplicate it if so desired.

As frontal dynamics is a time-dependent nonlinear mathematical problem, it is a challenging topic to learn and to teach. It would be a daunting task for instructors to prepare comprehensive and rigorous teaching materials on frontal dynamics. This book is especially intended for them as well as for the highly motivated student who wishes to learn about frontal dynamics alone. An abstract of each chapter of this book is given in Section 1.6. Sections 2.1–2.3, 3.2, 4.1 and 4.2, cover the rudimentary aspects of frontal dynamics. These materials, together with Chapter 1, would be suitable for an undergraduate course. In addition, one could select among the materials in Sections 2.4, 2.5, 3.3, 4.3 and 4.4, and Chapters 5, 6, and 7 for a graduate course.

Mankin Mak

Acknowledgments

The feedback of Professors Paul Markowski and Peter Bannon at the Pennsylvania State University in the early stage of writing this book has been valuable and much appreciated. Mr. Matt Lloyd, the editor of Cambridge University Press, has been very helpful from the start to the finish of this project. I wish to thank Professors Norman A. Phillips, Ming Cai and Yi Deng for their valuable comments on the draft of Chapters 1, 3, and 6, respectively. Professor Larry Di Girolamo and Dr. Guangyu Zhao at the University of Illinois kindly provided the digital data for the satellite image appearing on the front cover of this book. That figure, containing supplemental structural properties of a mid-latitude cyclone, is taken from one of two very recent research articles coauthored with Mr. Yi Lu and Professor Yi Deng at the Georgia Institute of Technology. The findings in those articles serve as the basis for my discussion of the general dynamics of atmospheric fronts in the context of a state-of-the-art model. Finally, I would like to gratefully acknowledge the permission granted by the American Meteorological Society and the Royal Meteorological Society to use many figures from their journal articles in this book.

1

Overview

1.1 What Are Fronts Like?

What fronts are like and why they are as we observe them to be are the two questions this book is all about. Forecasters in mid-latitude arguably pay more attention to the current whereabouts of the fronts and their expected evolution than to other elements of the atmosphere when they make short-range weather predictions. The American Meteorological Society (AMS) gives a definition of a front in the Glossary of Meteorology as: "the interface or transition zone between two air masses of different density." It follows that meteorological variables, such as temperature, wind, and moisture, would vary across a front by a much larger amount than they do so along it. One main category of fronts is known as surface fronts (SFs), existing next to the Earth's surface. There is, however, another class of fronts intimately related to the tropopause, known as upper-level fronts (ULFs). The AMS glossary also states that "an upper front is present in the upper air but typically does not extend to the ground." It reflects a commonly held notion that SFs and ULFs seldom interact. This statement makes no distinction between them other than their locations. They are actually quite different and may interact more often than one might expect.

Fronts at the surface have two main subcategories: *cold and warm fronts*. They are integral parts of an extratropical cyclone. Their properties are well established as we have had dense surface data for a long time. A cold front is labeled, by convention, on surface weather maps as a line laced with small blue triangles pointing in its direction of propagation. Severe weather elements embedded in meso-scale line segments typically flare up along and/or a short distance ahead of an advancing mature cold front. The severe weather outbreak associated with a cold front may consist of squalls and precipitation of all sorts accompanied by thunder and lightning, and even tornadoes. The mean wind strongly shifts across it in an anticlockwise direction about the low-pressure center of the cyclone in the northern hemisphere.

There is typically a much less intense, shorter, and wider *warm front* to the east of the surface low-pressure center. A warm front is indicated on surface weather

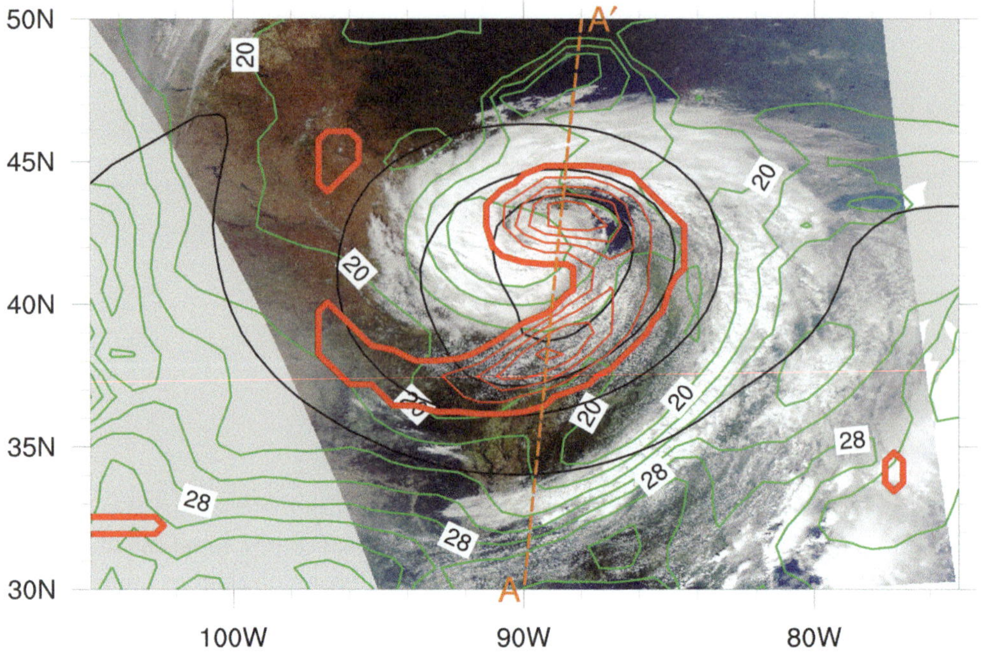

Figure 1.1. MODIS satellite image of a cyclone over the Midwestern United States at 1905Z on 26 September 2011. Three overlaid fields: (1) Temperature at the 2 m level (green contours, T_{min} = 13 °C, CI = 2 °C); (2) potential vorticity at 425 hPa (red contours, q_{max} = 4.6 PVU, CI = 1, thick red contour PV = 1, 1 PVU = 10^{-6} km^2 s^{-1} kg^{-1}); and (3) height of 425 hPa (black contours, Z_{min} = 6300 m, CI = 100 m), ERA-Interim reanalysis data at 1800Z.

maps by a line laced with small red semi-circles pointing in its direction of propagation. Moderate but more lasting precipitation usually spreads over a warm front region. It is mostly associated with stratus and stratocumulus clouds. The line segment between a low-pressure center and the intersection of a cold front and a warm front is referred to as an *occluded front*. It would be indicated on surface weather maps by a line segment laced alternately with small triangles and semi-circles. A line with symbols alternately pointing in opposite directions would indicate a front that moves very slowly. It is naturally referred to as a *stationary front*. All of these are collectively referred to as SFs.

The properties of ULFs are more subtle and require considerable effort to detect and document. A single figure showing both SF and ULF coexisting in a particular mid-latitude cyclone would give us a broad view of what fronts are like. Figure 1.1 shows a satellite image of MODIS (Moderate Resolution Imaging Spectroradiometer) taken at 1905Z on 26 September 2011, when it flew over the Midwestern United States. This disturbance has a gigantic comma-shaped cloud field made up of several types of cloud in different parts of it: innumerable low-level shallow cumulous clouds; organized vigorous convective cloud bands; and high-level cirrus cloud decks. The comma pattern of clouds has a signature "tail" in a northeast–southwest orientation of about 1500 km in length. Overlaid onto this satellite image are contours of three meteorological variables constructed with the use of the ERA-Interim (ECMWF reanalysis dataset) at a slightly earlier time (1800Z). One of these fields is "two-meter temperature" (green contours). It increases by about 10°C over a distance of 100 km across the most intense part of the signature "tail" in the cloud field. The curvilinear line at the western edge of it is a *cold front* separating surface warm moist air to the southeast from cold dry air to the northwest.

Figure 1.1 also shows the distributions of geopotential height (Z, black contours) and potential vorticity (PV, red contours; to be mathematically defined later) at 425 hPa level. The lowest contour of Z is 6300 m, practically overhead of the minimum temperature center at the surface. In other words, this disturbance has developed an upper-level cutoff low and is in its mature stage by this time. PV and potential temperature are quasi-conservative properties of air parcels if diabatic heating and turbulent mixing are both negligible. The PV in the stratosphere of the mean atmosphere is much larger than in the troposphere, considerably larger than $1\ \text{PVC} = 10^{-6}\ \text{ks}^{-1}\ \text{kg}^{-1}\ \text{m}^2$. Suffice to note for the moment is that there is an extensive area in Figure 1.1 enclosed by a contour of 1 PVU. The boundary of this area is highlighted by the thick red contour. The maximum PV value within this enclosed area is equal to 4.6 PVU. The air in the enclosed area is therefore of stratospheric origin and must have been brought down to this tropospheric level through a *folding of the tropopause*. This inference is consistent with the fact that the locations of the two centers of maximum PV values are distinctly less cloudy than their immediate surroundings. The boundary of this area is therefore a demarcation separating tropospheric air from air of stratospheric origin. The thick red contour may be interpreted as a ULF, which has the shape of a closed loop at this level in this case. This ULF and the SFs are therefore integral parts of an extratropical cyclone, but have distinctly different properties.

Figure 1.2 is a complementary figure showing the vertical distribution of PV and potential temperature, θ, on a cross-section passing through the locations A (30N, 90W) and A′ (50N, 88W). This cross-section passes through the two maximum centers of PV on the 425 hPa surface. This figure gives us a glimpse of the *vertical structure* of both SF and ULF. Recall that the contour of PV = 1 PVU at the 425 hPa level is a closed loop. Here we see that it extends down to quite low levels in the troposphere at different locations. We may visualize this ULF as being somewhat like a drawn curtain. A location where the θ contours bunch up together in this cross-section is another meaningful indicator of the presence of ULF. The southern portion of this ULF is more pronounced and extends to lower levels than the northern portion in this cross-section. It is noteworthy that the steeply sloping θ contours extend all the way down to the surface near (30N, 90W). This figure suggests that the shallow surface cold front appears to be linking up with the ULF aloft. Sometimes, two such fronts could dynamically interact when they are favorably

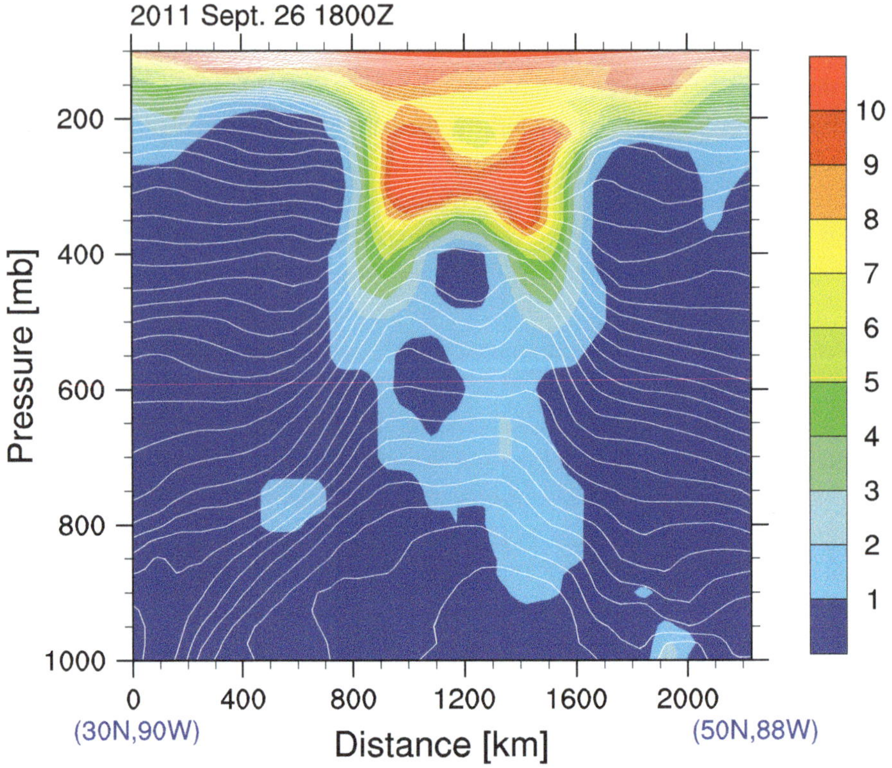

Figure 1.2. Distribution of potential vorticity in PVU (shading) and potential temperature (contours, CI = 2 K) on a vertical cross-section passing through the points A (30N, 90W) and A′ (50N, 88W) at 1800Z on 26 September 2011.

located relative to one another. Much more about the frontal system in this extratropical cyclone and the related dynamics will be discussed in Section 7.7.

There are fronts in the oceans too (von Arx et al., 1955; Fedorov, 1984). They are much shorter than those in the atmosphere as the oceanic eddies are much smaller than the counterpart atmospheric cyclones. Thus, frontal dynamics is a well-defined general aspect of geophysical fluid dynamics. Measuring the properties of oceanic fronts is a costly and challenging task for it would require specially designed instruments to be launched by seagoing ships for an extended period of time. It is reassuring that fronts can also be subjected to laboratory investigation with a rotating dishpan (Fultz, 1952; Faller, 1956), although making quantitative measurement is difficult. In this book, unless explicitly stated otherwise, it is understood that atmospheric fronts are simply referred to as "fronts."

Having seen the salient features of fronts in this section, one would naturally wonder: What are the other important aspects in the structures of cold and warm surface fronts? How fundamentally different is ULF from SF? Where do fronts originate? How and why do fronts form, move, evolve, and eventually disappear? Might SF and ULF even synergistically interact?

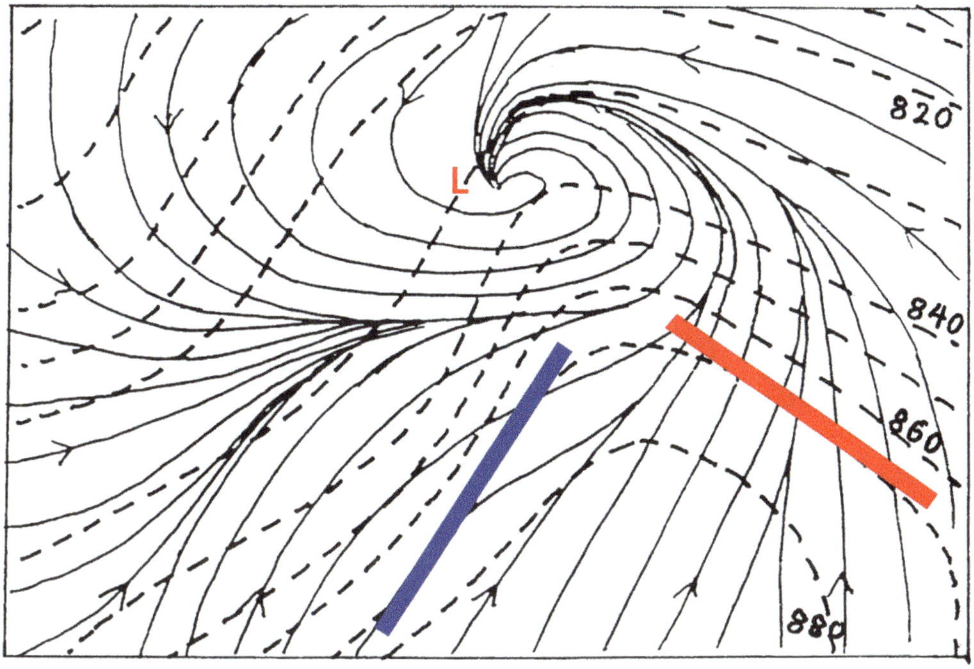

Figure 1.3. The 1000–900 mb thickness map (intervals 10 dynamic meters – 3 K) superimposed upon the streamlines at ground level; adapted from Bjerknes et al. (1911).

1.2 Observational Studies of Some Historical Significance

A century ago meteorologists were only dimly aware of the existence of atmospheric fronts. There are two diagrams of special historic interest. They highlight the degree of success in the early studies of fronts. Bjerknes et al. (1911) published a hand-made surface weather map with minimal labeling (Figure 1.3). It seems to have enabled Vilhelm Bjerknes to get multi-year funding for his budding research endeavor in Norway from the Carnegie Institute of Washington in the United States. It is arguably the oldest research weather map constructed, with only a meager amount of surface meteorological data in North America. Nevertheless, it correctly depicts the surface wind distribution in a synoptic-scale disturbance and hints at the existence of fronts on 28 November 1905. It shows a surface flow cyclonically spiraling toward a certain location in Minnesota, United States. A label "L" is added in the map next to that locality to indicate a low-pressure center. The solid contours are streamlines of the flow, thereby depicting a cyclonic flow around a low-pressure center. The dashed contours are the thickness lines of the 1000–900 mb layer in meters. Those contours effectively depict the temperature field by virtue of hydrostatic balance. Therefore, it would be justifiable for us to add a blue line to highlight the presence of a cold front to the south of the "L" and a red line to highlight the presence of a warm front to the southeast of "L." Those two lines have

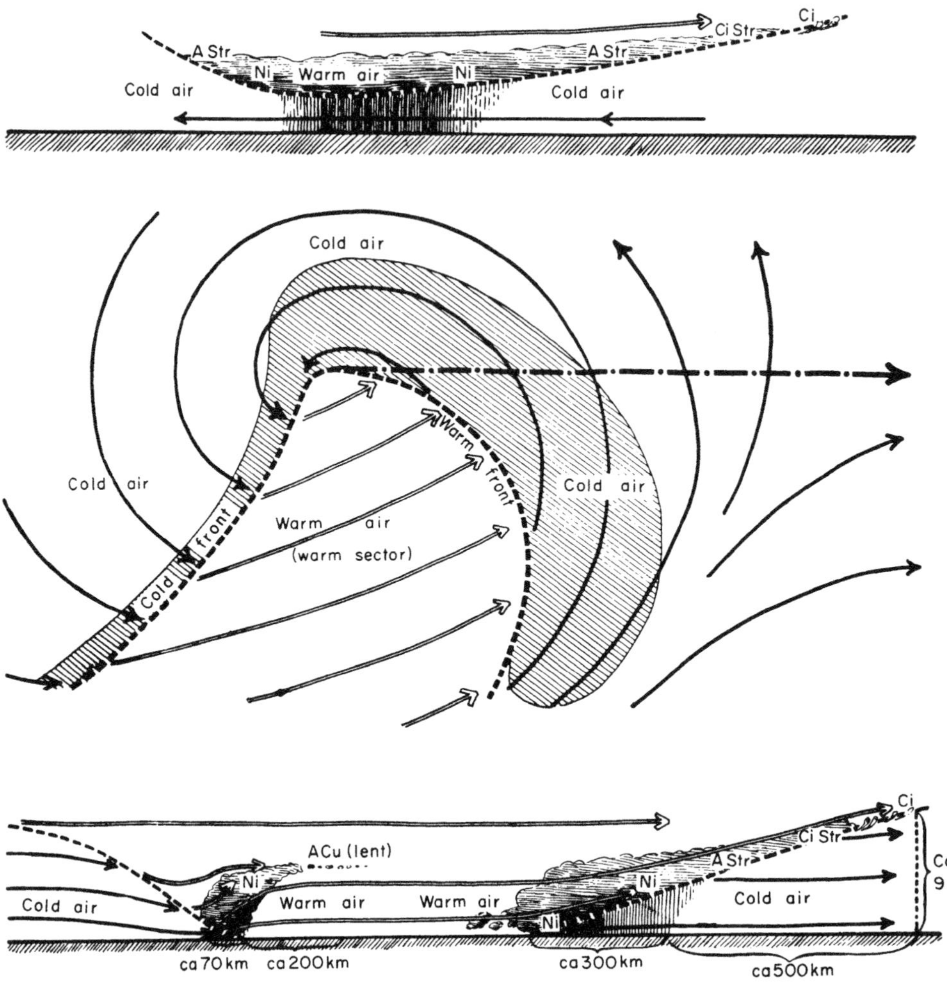

Figure 1.4. Idealized cyclone, from Bjerknes and Solberg (1921). In the middle diagram, dash-dotted arrows show the direction of motion of the cyclone; other arrows are streamlines of airflow at the Earth's surface. Top and bottom diagrams show the cloud system and air motions in vertical sections along the direction of cyclone movement north of its center and across the warm sector south of its center.

been added to the figure. These lines exemplify what Vilhelm Bjerknes metaphorically called *"wrinkles in the weather's face"* (Jewell, 1994).

The construction of Figure 1.4 by Bjerknes and Solberg (1921) heralded a systematic research program by a small band of young Norwegian meteorologists under the leadership of Vilhelm Bjerknes (later known as the Bergen School). The focus of their work was delineating the physical nature of cyclones, fronts, and related weather elements. The capstone of their work is a historically important schematic diagram of atmospheric fronts in relation to a mid-latitude cyclone. This schematic synthesizes the surface observational information about the weather elements within the single synoptic-scale disturbance in mid-latitude known at that time. It depicts a general cyclonic flow pattern around an eastward-moving low-pressure center. The configuration of the surface flow in Figure 1.4 is much more realistic than that in Figure 1.3. Not only does it correctly identify the typical positions of a narrow cold front together with a substantially broader warm front region with respect to the low-pressure center, but also highlights the weather elements associated with them. The cold front and warm fronts are shown to emanate from a low-pressure center. Cumulonimbus clouds are indicated over the cold front, with attendant strong updraft. A much more gentle airflow gliding over the warm front gives rise to a layer of stratocumulus clouds with an extensive deck of cirrus aloft in the downstream direction. Precipitation of all sorts falls in both frontal regions. However, no occluded front is indicated in this schematic. This schematic is a graphic portrait of the so-called *Polar Front Theory*. For many years, this theory served as a field guide for weather forecasters and a qualitative paradigm for SFs.

It was not until well after the Second World War that comprehensive observational diagnoses of ULF could be performed using routine upper-air data in conjunction with research aircraft data. A good example of such studies is shown in Figure 1.5 (Shapiro et al., 1984). Panel (a) shows a deep trough in the 300 mb flow at 12Z on 13 June 1983. Panel (b) shows the distribution of potential temperature as well as the isotach of the wind normal to a cross-section along the dotted line AA′ in panel (a) through the western jet. Visual inspection of panel (b) suggests there is an important exchange of stratospheric air associated with the ULF. The construction of Figure 1.5 demonstrates the capability of the current data available for frontal research.

1.3 Frontogenetical Function for Surface Fronts

A surface front is characterized by its highly localized large magnitude in the horizontal gradient of potential temperature, $|\nabla_2\theta|$. To measure its rate of evolution as it moves around, it would be reasonable to use $F_2 \equiv \frac{D}{Dt}|\nabla_2\theta|$ as a metric where $\frac{D}{Dt} = \frac{\partial}{\partial t} + \boldsymbol{V}\cdot\nabla$ is the total derivative, $\boldsymbol{V} = (u, v, w)$ and $\nabla = \left(\frac{\partial}{\partial x}, \frac{\partial}{\partial y}, \frac{\partial}{\partial z}\right)$ in an (x, y, z) coordinate system. With the use of the thermodynamic equation, the explicit expression of F_2 can be written as

$$\frac{D}{Dt}|\nabla_2\theta| = \frac{1}{|\nabla\theta|}\left\{\theta_x\left[\frac{1}{c_p}\left(\frac{p_o}{p}\right)^{\kappa}\frac{\partial H}{\partial x} - u_x\theta_x - v_x\theta_y - w_x\theta_z\right]\right.$$

$$\left.+\theta_y\left[\frac{1}{c_p}\left(\frac{p_o}{p}\right)^{\kappa}\frac{\partial H}{\partial y} - u_y\theta_x - v_y\theta_y - w_y\theta_z\right]\right\}, \tag{1.1}$$

where H stands for the net diabatic heating rate including the conversion from kinetic to thermal energy by turbulent mixing, $c_p = 1004\ \mathrm{J(kg)^{-1}\ K^{-1}}$, p the

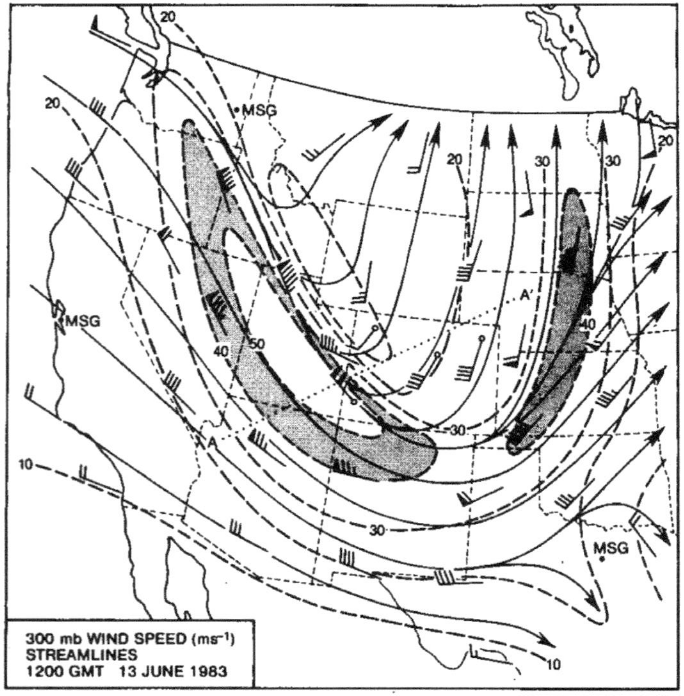

(A)

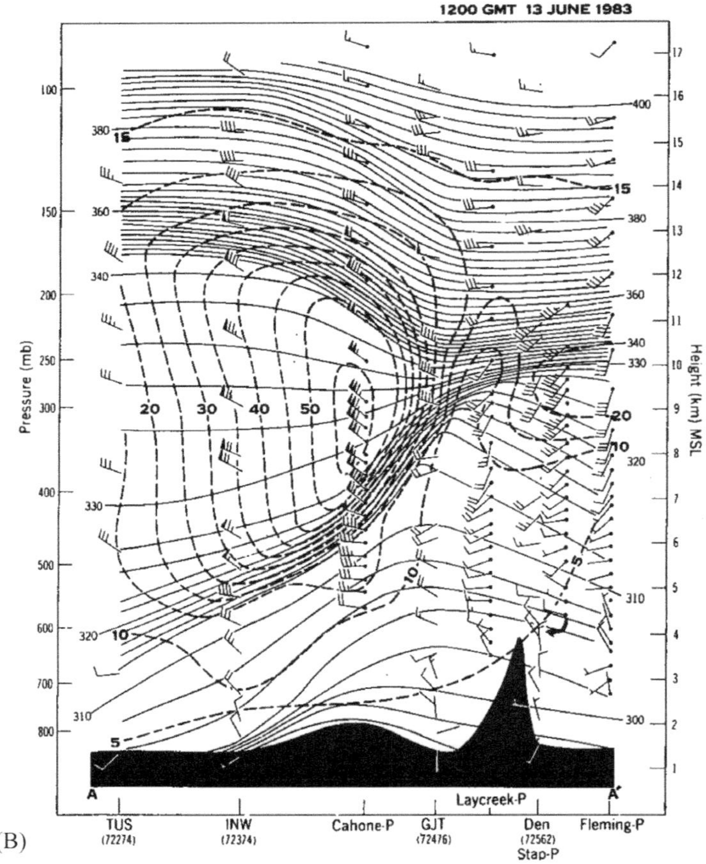

(B)

Figure 1.5. (a) Wind speed and streamlines at 300 mb on 1200Z 13 June 1983. (b) Distribution of potential temperature and wind speed in the cross-section in panel (a), from Shapiro et al. (1984).

pressure, p_o = 1000 hPa, a reference pressure, $\kappa = \frac{R}{c_p} = 0.286$ with R being the gas constant for dry air. The remaining terms on the RHS of (1.1) stand for the effects of the flow field acting on the temperature field. We call this definition the "frontogenetical function on x–y plane" for short.

The dynamic terms are traditionally discussed in terms of the following two groupings:

$$\frac{1}{|\nabla_2\theta|}[-\nabla_2\theta\cdot\boldsymbol{Q}_{total}] \quad \text{and} \quad \frac{1}{|\nabla_2\theta|}[-\theta_z\,(\nabla_2 w\cdot\nabla_2\theta)], \tag{1.2}$$

where $\boldsymbol{Q}_{total} = (\nabla_2 u\cdot\nabla_2\theta,\ \nabla_2 v\cdot\nabla_2\theta)$. The first group of terms is referred to as the influence of the deformation property of the total horizontal wind since it can be rewritten as

$$-\frac{1}{|\nabla_2\theta|}\nabla_2\theta\cdot\boldsymbol{Q}_{total} = -\frac{1}{|\nabla_2\theta|}\frac{1}{2}\left(\theta_x^2-\theta_y^2\right)D_1 - \frac{1}{|\nabla_2\theta|}\theta_x\theta_y D_2, \tag{1.3}$$

where $D_1 = u_x - v_y$ is stretching deformation and $D_2 = u_y + v_x$ is shearing deformation. This expression tells us that the stretching deformation of the flow field, D_1, would not contribute at the locations where the isotherms are roughly in the SW–NE direction $\left(|\theta_x|\sim|\theta_y|\right)$. On the other hand, the shearing deformation of the flow field, D_2, would not contribute at the locations where the isotherms are either close to the zonal direction ($\theta_x \sim 0$) or meridional direction ($\theta_y \sim 0$). A horizontal wind that is characterized by stretching deformation would tend to squeeze the isotherms toward a certain direction, known as an axis of dilatation. A horizontal wind that is characterized by shearing deformation would tend to align and tighten the isotherms ever closer together by the action of the shear.

The second group of terms in (1.2) measures the influence arising from the horizontal variation of the vertical velocity. A common terminology for such a term is the so-called "tilting" effect. This is similar to the action of shearing deformation, except by a flow on a vertical plane. One can readily visualize these three processes in action. There are schematic diagrams in most textbooks that serve to graphically depict these processes in action, leading to the development of a surface front.

We may take an alternative viewpoint by examining the influence of a flow field in terms of its geostrophic and ageostrophic components. In order to reserve the use of subscripts exclusively for denoting partial derivatives, we choose not to use the commonly used subscripts "g" and "a" for indicating geostrophic and ageostrophic wind components. Instead, we use "^" and "~" respectively for that purpose. Thus, we decompose the horizontal velocity as: $u = \hat{u} + \tilde{u}$, $v = \hat{v} + \tilde{v}$ with $(\hat{u}, \hat{v})$ being the geostrophic wind, $(\tilde{u}, \tilde{v})$ the ageostrophic wind. $(\tilde{u}, \tilde{v}, \tilde{w})$ is a 3D ageostrophic velocity vector. With these notations, we can rewrite (1.1) as

$$\frac{D}{Dt}|\nabla_2\theta| = \frac{1}{c_p|\nabla_2\theta|}\nabla_2\theta\cdot\nabla_2\left(\left(\frac{p_o}{p}\right)^{\kappa}H\right) - \frac{1}{|\nabla_2\theta|}\nabla_2\theta\cdot\boldsymbol{Q} - \frac{1}{|\nabla_2\theta|}\nabla\theta\cdot\boldsymbol{B}, \tag{1.4}$$

where $\boldsymbol{Q}$ is

$$\boldsymbol{Q} = (\nabla_2 \hat{u} \cdot \nabla_2 \theta, \quad \nabla_2 \hat{v} \cdot \nabla_2 \theta) \tag{1.5}$$

and

$$\boldsymbol{B} = (\nabla_2 \tilde{u} \cdot \nabla_2 \theta, \quad \nabla_2 \tilde{v} \cdot \nabla_2 \theta, \quad \nabla_2 \tilde{w} \cdot \nabla_2 \theta). \tag{1.6}$$

A symbolic form of (1.4) is

$$F_{2T} = F_{2H} + F_{2G} + F_{2A}, \tag{1.7}$$

where

$$\begin{aligned} F_{2H} &= \frac{1}{|\nabla_2 \theta|} \nabla_2 \theta \cdot \nabla_2 \left(\frac{H}{c_p} \left(\frac{p_o}{p} \right)^{\kappa} \right) \\ F_{2G} &= -\frac{1}{|\nabla_2 \theta|} \nabla_2 \theta \cdot \boldsymbol{Q} \end{aligned}$$

and

$$F_{2A} = -\frac{1}{|\nabla_2 \theta|} \nabla \theta \cdot \boldsymbol{B}.$$

F_{2H} is the effect of net diabatic heating on the rate of frontogenesis. F_{2G} and F_{2A} are measures of the dynamical effect of the geostrophic component and the 3D ageostrophic component of the flow on frontogenesis, respectively. Equation (1.4) will be verified to be adequate for measuring the development of surface fronts in Section 7.5. But we have reasons to suspect it not to be the case for measuring ULF. We will formulate a more general frontogenetical function and illustrate it to be appropriate for measuring the development of ULF in Section 7.6.

It should be emphasized that identifying the physical meaning of the individual terms on the RHS of (1.4), as well as knowing their quantitative values for a specific front, does not constitute a sufficient understanding of frontogenesis. While the expression of frontogenetic function serves as a useful diagnostic tool, it does not tell us the causal relationships among the variables in a prognostic sense. In fact, only the thermodynamic equation has been invoked in the derivation of (1.4) and the law of motion is only implicitly underlying the relationship among all the dependent variables. An understanding of frontal dynamics must be built upon our ability to simulate the formation and evolution of front(s) under the simultaneous influences of the physical processes of first-order importance in a model framework. A simulation should start from an initial state that has no frontal characteristics at all. In other words, it is necessary to construct and analyze self-contained mathematical frontal models for specific objectives. What needs to be done will be greatly elaborated with a number of illustrative analyses in the remaining chapters of this book.

1.3.1 The First Diagnosis of F_2

Sanders (1955) performed the first diagnosis of frontogenetical function for an intense surface frontal zone using radiosonde and upper-level wind stations in the south-central United States on 17 and 18 April 1953. The front was treated as if it were unidirectional parallel to an x-axis, along which all dependent variables were uniform. The corresponding frontogenetical function is then $\frac{D}{Dt}|\theta_y|$. There were no data for estimating the diabatic heating rate. Eq. (1.1) is therefore reduced to

$$\frac{D}{Dt}|\theta_y| = -\frac{\theta_y}{|\theta_y|}\{v_y\theta_y + w_y\theta_z\} \equiv F_{2A}, \tag{1.8}$$

where $v = \tilde{v}$ and $w = \tilde{w}$. With the x-axis pointing north in the case of a cold front, the y-axis would point toward the cold sector. So, we would have $\frac{\theta_y}{|\theta_y|} = -1$. Sanders' results of the individual parts of F_{2A} are shown in Figure 1.6. It reveals that the confluence term has large positive values and the tilting term has negative values in this frontal zone near the surface. The two terms cancel in the frontal zone. The tilting term has significant positive values immediately ahead of the cold front, signifying a tendency for eastward movement of the front. In light of the quality of the data at that time, we should only expect such results to be suggestive.

1.4 Literature Review of Frontal Dynamics

Bergeron (1928) pointed out that a wind field with significant stretching deformation would be a favorable factor for development of fronts. It could progressively realign the contours of any material conservative property, such as potential temperature, toward the axis of dilatation and tighten the spacing between them. Apart from tightening up the isotherms, the wind field would also generate vorticity in the vertical direction via vortex stretching and hence induce horizontal variation in the vertical motion field. These processes are indicated by the dynamic terms on the RHS of (1.1) as a measure of the developmental rate of an idealized straight front. There were many diagnostic studies (e.g., Newton 1954; Sanders 1955; Reed and Danielson 1959; Keyser and Shapiro, 1986) that reported the assessment of such frontogenetical function.

Side-stepping the issue concerning the origin of fronts, the Bergen School in the early decades of the twentieth century postulated that cyclones would emerge from a preexisting front by virtue of some form of hydrodynamic instability. By implication, the cold front and warm front embedded in a cyclone would be merely modified versions of the preexisting front. Their attempts to validate that postulate were, however, unsuccessful since such unstable waves have quite different structural characteristics from those of the mid-latitude cyclones. Observations suggest that extratropical cyclones typically emerge without a preexisting front and that the surface temperature gradient in certain locations within the cyclone progressively and

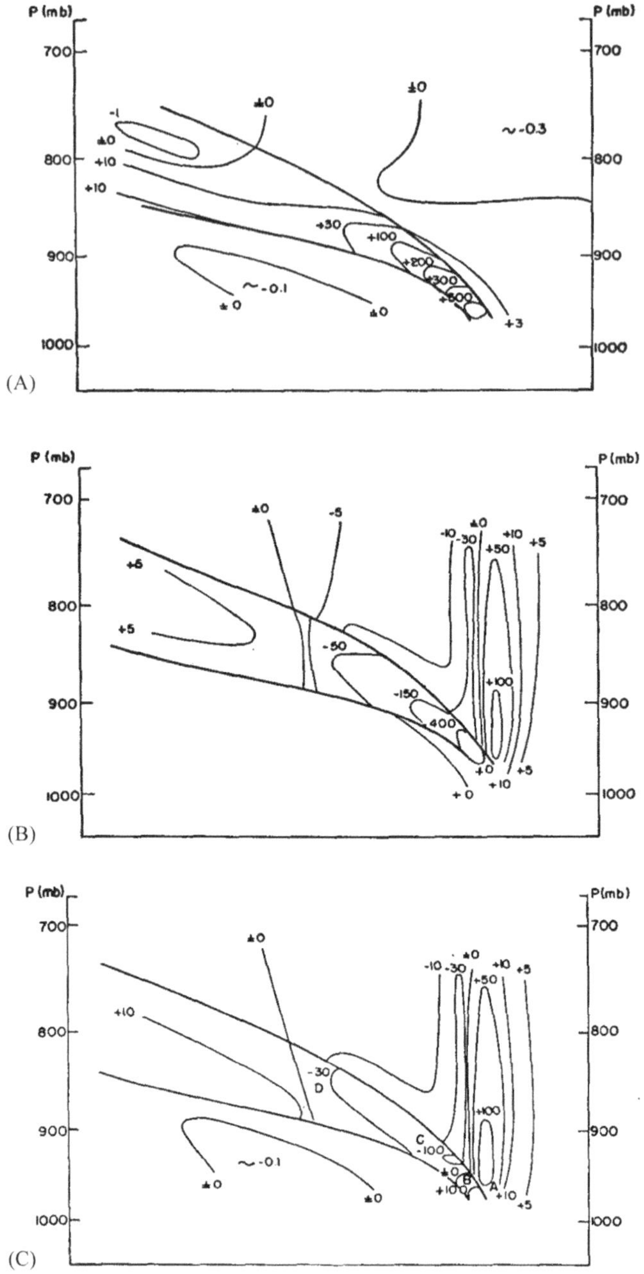

Figure 1.6. Distribution of (a) $\frac{\partial v}{\partial y}\frac{\partial \theta}{\partial y}$, (b) $\frac{\partial w}{\partial y}\frac{\partial \theta}{\partial z}$ and (c) their sum. Units in C/(3 h 100 km) on a cross-section normal to a cold front (taken from Sanders, 1955).

greatly increases (hence frontal formation) during its intensification. Therefore, the opposite of the postulate of the Bergen School turns out to be closer to the truth in that *cyclogenesis gives rise to frontogenesis* rather than vice versa. In other words, frontogenesis is intrinsically a time-dependent nonlinear dynamical problem and cyclones do not emerge from instability of a front. A satisfactory theory of cyclogenesis is then a prerequisite for establishing an adequate theory of frontogenesis. That is why frontogenesis is a more difficult problem than cyclogenesis.

A generic theory of cyclogenesis was independently established by Charney (1947) and Eady (1949). By systematically invoking the *quasi-geostrophic (QG) approximation*, they reduced the complete set of governing equations in their respective atmospheric models to a single prognostic equation: the QG potential vorticity equation. They worked out the analytic solutions of their respective linearized QG potential vorticity equation about a simple basic baroclinic state. Such solutions represent wave disturbances that have structural properties broadly reminiscent of the observed mid-latitude cyclones, although the two models differ in certain important aspects. They therefore demonstrate that formation of an extratropical cyclone is understandable from the perspective of *baroclinic instability* of the flow in a background baroclinic zone. Charney referred to such unstable waves as "long upper-air waves" in distinction from the "frontal waves" near the surface. These two classes of waves differ in periods (days versus hours) and wavelengths (6000 km versus 1000 km). Charney specifically remarked in his paper that his "investigation does not tell what relationship exists, in the generation of cyclones, between the frontal perturbation and the long atmospheric waves" but conjectured that "there must be a linkage between the frontal and long waves."

The first investigation of cyclogenesis in finite amplitude was performed by Phillips (1956) with a two-layer QG dry model, albeit in the context of a numerical simulation of the general circulation of the atmosphere. Limited by the computer technology at that time, the model had very low resolution. Referring to the results presented in his figures, Phillips stated in his paper: "Definite indications of something similar to cold and warm fronts are to be seen in the 1000-mb contours, with the main temperature gradient occurring on the cold side of the 'frontal' troughs." It was attributed to the impact of the deformation flow in the developing model cyclones and anticyclones. This finding is the first quantitative evidence in record suggesting that frontogenesis could arise from cyclogenesis and not vice versa. However, a proto-front in a QG model such as Phillips' is not a good simile of the real atmospheric fronts. It intrinsically could not develop to great intensity with a width comparably as narrow as the observed fronts (even in high resolution and without dissipation). The intrinsic limitation stems from the nature of the QG approximation. The related omega equation governing the ageostrophic flow component would remain elliptic at all times, lacking the indispensable feedback process for generating intense fronts. Alas, it became increasingly evident in the 1960s that even a high-resolution QG model (e.g., Williams, 1968; Williams and Plotkin, 1968) is intrinsically incapable of simulating bona fide fronts.

Eliassen (1948) pointed out that under certain conditions it would be justifiable and better to approximate the acceleration of the horizontal velocity by the acceleration of the geostrophic flow component without approximating the total advection by the three velocity components. That mathematical treatment is known as *geostrophic momentum approximation*, which is considerably less restrictive than the QG approximation. A model that explicitly invokes such approximation is known as a *semi-geostrophic (SG) model*. The key to deriving the corresponding simplified model equations with a method of singular perturbation expansion is to take into consideration the coexistence of two different horizontal spatial scales in a disturbance with fronts embedded in it. The simplest version of an SG model is one in a two-dimensional setting. An integral part of an analysis of the flow in a 2D SG model is to compute the transverse (ageostrophic) circulation in a cross-section normal to a front at each time instant as a functional of the geostrophic circulation in the along-front direction. The equation that governs such transverse circulation is known as the Sawyer–Eliassen (SE) equation (Sawyer, 1956; Eliassen, 1962). It is an extension of the QG omega equation. The SE equation would not necessarily remain an elliptic partial differential equation during the evolution of a cyclone in an SG model.

The breakthrough in the study of frontal dynamics was achieved by Hoskins and Bretherton (1972) with a two-dimensional SG model on the f-plane. By making use of the so-called *geostrophic coordinates* they obtained an analytic solution for the nonlinear evolution of an unstable baroclinic wave in the presence of an initial Eady shear flow in a 2D SG model. The SE equation would take on the same form as the QG omega equation when it is expressed in the geostrophic coordinates. Their result analytically demonstrates that frontogenesis is indeed a natural byproduct of cyclogenesis under the joint influences of the simultaneously intensifying geostrophic and ageostrophic flow components. *Their model front evolves toward a singularity in finite time.* That is the time when the SE equation breaks down in the sense that it no longer remains an elliptic partial differential equation and therefore no solution for the boundary value problem could exist. This would occur when the shear near the surface has become too strong and/or the stratification at some locations has become statically unstable. Those are the feedback influences of the ageostrophic circulation itself. Transient surface frontogenesis is also possible in this model setting, even when the initial disturbance has a wavelength shorter than the short-wave cutoff for baroclinic instability, provided that it has a nonzero potential vorticity anomaly capable of interacting with the surface baroclinicity, albeit only for a finite duration (Montgomery and Farrell, 1990).

There are studies suggesting that friction should have a significant impact on the structure of surface fronts by limiting the minimum width and maximum intensity of a front (e.g., Eliassen, 1962; Blumen and Wu, 1982; Cooper et al., 1992; Doswell and Haugland, 2007). There are quite a number of articles in the literature that especially look into the impact of surface friction on the structure of SFs embedded in a 2D baroclinic wave model. Such a model, however, intrinsically cannot examine the differences between cold and warm fronts. In a more general study, Hines and Mechoso (1993) experiment with different values for the drag coefficient in a 3D adiabatic primitive equation model. They find "a distinct warm front does not develop in the simulation with a surface drag coefficient representative of continental surfaces. Cold frontogenesis, on the other hand is not sensitive to surface drag." They interpret their finding by suggesting that "[while] surface drag weakens the warm advection and slows the intensification by positive feedback … cold advection-type frontogenesis … does not appear to include the drag-sensitive positive feedback." Such a result of the warm front simulation is somewhat puzzling/problematic since distinct warm fronts do frequently form in the atmosphere over continental surfaces. Furthermore, warm fronts and cold fronts do simultaneously form together over a given land surface, such as North America. Sinclair and Keyser (2015) compare the balance of forces in an idealized front in the general model using different horizontal grid resolutions with versus without a planetary boundary layer parameterization scheme. Their key finding is that unbalanced flow would develop at the leading edge of the cold front only when very high resolution is used (4 km grid spacing). With dissipation, frontal collapse would be naturally suppressed in an SG model. There would be damped oscillations after the model front has reached its maximum intensity (Nakamura, 1994). It is reasonable to believe that friction has only quantitative but not qualitative impacts on frontogenesis.

There have been attempts to quantitatively examine the impacts of condensational heating on surface frontogenesis initiated with an unstable mode in 2D SG models using different parameterization schemes (e.g., Mak and Bannon, 1984; Thorpe and Nash, 1984; Emanuel et al., 1987; Davies and Muller, 1988). Simple treatments of such heating were found to have strong but not qualitatively different impacts upon frontogenesis. Condensational heating was also found to considerably strengthen transient surface frontogenesis in a 2D SG model (Montgomery and Farrell, 1991).

Instructive as a 2D SG model is, the constraint of two-dimensionality in such a model proves to be too restrictive. In particular, a 2D SG model cannot dynamically distinguish between cold and warm fronts. It is more suitable for simulating the development of a cold front than of a warm front. The underlying dynamical reason for it is that the 2D Eady model precludes some of the interactions of a disturbance with the mean flow.

The 3D inviscid adiabatic semi-geostrophic model has played a valuable pedagogical role in the studies of atmospheric fronts. It is the simplest model setting adequate for investigating the intrinsic dynamics of the life-cycle of fronts. There were a number of model studies by Hoskins and his collaborators (Hoskins, 1975, 1976; Hoskins and Draghici, 1977; Hoskins and West, 1979; Hoskins and Heckley, 1981; Schar and Wernli, 1993) that extensively applied this tool with the use of geostrophic coordinates first suggested by Yudin (1955):

$$X = x + f^{-1}v, \quad Y = y - f^{-1}u, \quad Z = z, \quad T = t, \tag{1.9}$$

where u and v are the unknown geostrophic velocity components. These coordinates have the advantage of transforming the set of governing equations that incorporate the geostrophic momentum approximation similar *in form* to those of a quasi-geostrophic model. With additional approximation they would become identical. The underlying reason is that the coordinate transformation continually stretches a locality of cyclonic vorticity and compresses a locality of anticyclonic vorticity while nonlinear dynamics induces a progressive contraction of scale across a frontal zone. The self-stretching coordinates would disguise the presence of the higher harmonics, and thereby would only implicitly represent the dynamical role of the ageostrophic circulation. However, the advantage gained in using the geostrophic coordinates does not come without a hefty price, since the dependent variables would have to be transformed to highly convoluted forms (e.g., eq. (10) in Hoskins and Draghici, 1977) Furthermore, the procedure for obtaining the geostrophic streamfunction from an updated field of SG potential vorticity via inversion is often simplified by neglecting a supposedly small nonlinear term. The latter, however, understandably becomes progressively more significant as a front intensifies.

It is actually much simpler to analyze frontal development in physical coordinates (x, y, z) (i.e., not using the geostrophic coordinates). We can do so without imposing additional restriction. Mak (2014) has recently reported such an analysis of the intrinsic dynamics of a life-cycle of SFs. To obtain the 3D ageostrophic circulation in this setting, he generalized the SE equation to two coupled partial differential equations that govern two scalar functions for representing the ageostrophic circulation. The latter can then be determined at each instant as a functional of the geostrophic streamfunction. The SG model simultaneously simulates cold and warm fronts in physical coordinates, starting from an initial 3D unstable baroclinic wave disturbance that has no frontal structure whatsoever under the influence of a baroclinic jet. It delineates the nature of dynamical difference between cold and warm fronts.

As noted earlier, the ULF is a distinctly different class of front. Broadly speaking, a ULF forms when the tropopause locally folds during strong intensification of a synoptic baroclinic wave. Through such folding of the tropopause there would be large local exchange of stratospheric and tropospheric air. Thus, ULFs separate air of stratospheric and tropospheric origins (Reed, 1955; Newton, 1958; Reed and Danielson, 1959; Danielson, 1968). There are also reports of clear air turbulence in ULF, presumably associated with breaking gravity waves. Making use of research aircraft data together with radiosonde data, Shapiro (1978) resolved the 100 km cross-front scale of a ULF. He attributed the frontal structure to the forcing of an indirect transverse circulation by the geostrophic flow component in an upper-level jet-streak migrating through a background baroclinic wave (Shapiro, 1981; Shapiro et al., 1984). A 2D SG model was used by Hoskins (1972) to demonstrate the formation of a ULF driven by an imposed vertically uniform confluent flow. Keyser and Pecnick (1985a, 1985b) extended such investigation with a 2D primitive equation model using a nonmodal initial disturbance and external forcing. It was found that when the prescribed geostrophic flow has cold advection on the cross-front plane, the simulated ULF is more intense because the secondary circulation is more skewed than the counterpart in the case of warm advection. Heckley and Hoskins (1982) further used a 3D SG model in geostrophic coordinates to simulate a ULF. It is shown to be an integral part of an intensifying unstable baroclinic wave, albeit the model front was rather weak. Keyser and Shapiro (1986) wrote a review of the literature of ULF preceding 1986.

There have been case studies that report significant interaction between an upper-level jet-streak and a low-level jet-streak (e.g., Uccellini et al., 1985 and the many observational references cited therein). Martin et al. (1993) attribute the development of vigorous convection in a storm on the mid-Atlantic coast to a vertical superposition of two frontal circulations. Equally pertinent to note is that while two jet-streaks may interact, their related ULF and SF may not (as highlighted by a schematic in Shapiro, 1982). In other words, interaction of two jet-streaks is not synonymous with interaction between ULF and SF. Hakim and Keyser (2001) report analytic Green function solutions of the SE equation in which all coefficients are presumed constants with subjectively placed forcing in the form of delta-functions. Those solutions show that a ULF and an SF may or may not interact. When they interact, they may do so negatively or positively, dependent upon the sign, strength, and locations of the constituent parts of the forcing. Whether or not a particular ULF and a particular SF in the atmosphere would interact is clearly an issue of interest from both theoretical and practical points of view.

A model less restrictive than an SG model is a primitive equation model in which internal gravity waves could exist. The dynamic instability of a front-like shear flow in a primitive equation model had been a topic of investigation for decades (various members of the Bergen School; Orslanki, 1968). There were studies of frontal simulation with a 2D primitive equation model (e.g., Williams, 1967; Orlanski and Ross, 1977; Nakamura and Held, 1989) as well as a 3D primitive equation model (e.g., Murdrick, 1974; Shapiro, 1975; Hines and Mechoso, 1991). In particular, the impact of the meridional shear of a baroclinic jet seems to be inadequately represented by the SG approximation and can be adequately represented only in a primitive equation model (Rotunno et al., 1994). The detailed nature of frontal collapse has been closely examined even with 2D non-hydrostatic frontal models (Snyder et al., 1993), suggesting a significant role of the internal gravity waves at the last stage of development. Two recent articles have closely examined the dynamics of upper-level frontogenesis and surface frontogenesis with a state-of-the-art 3D model with diabatic heating (Mak et al., 2016, 2018).

1.5 Objective of the Book

The brief review above indicates that frontal dynamics has been a topic of systematic investigations and much has been learned about it. Most of the well-known textbooks on atmospheric dynamics, however, do not cover the topic of frontal dynamics at all (e.g., Lindzen, 1990; Emmanuel, 1994; Salmon, 1998; Green, 2004; Vallis, 2006; McWilliams, 2006; Marshall and Plumb, 2007), perhaps because fronts are sort of a misfit – neither being a large-scale nor meso-scale disturbance. A number of textbooks give largely qualitative and kinematic discussions of fronts and frontogenesis supplemented by schematic diagrams (Palmen and Newton, 1969; Bluestein, 1993; Martin, 2006; Markowski and Richardson, 2010; Lackmann, 2011; Trapp, 2013). The discussions of frontal dynamics in others do not go beyond showing frontogenesis in a 2D SG model driven by an Eady flow or a specified confluent flow (Gill, 1982; Holton, 1992; Lin, 2008; Pedlosky, 1987; Mak, 2011; Hoskins and James, 2014). It also warrants mentioning that there is a distinctly different book steeped in discussing the mathematical nature of the SG equations (Cullen, 2006).

An undergraduate course of dynamic-synoptic meteorology typically covers only the kinematic of fronts and their rudimentary dynamics. It is unusual that frontal dynamics would be taught in depth even in a graduate curriculum of atmospheric sciences because it would require dwelling on nonlinear mathematical analyses. Another reason is perhaps that instructors, whose expertise is typically of one subfield or the other, would teach either large-scale dynamics or meso-scale dynamics. Since the front is a hybrid disturbance with characteristics of both large and meso scales, it suffers from the proverbial danger of "falling through the cracks" in a typical curriculum and does not receive the close attention it warrants. Due to the lack of textbook materials on frontal dynamics with self-contained illustrative analyses, it would be a daunting task for instructors to try preparing their own. The objective of this book is to provide instructors with materials on frontal dynamics illustrated with a number of self-contained analyses in a hierarchy of models of increasing complexity, so that motivated students might be able to learn the materials even by themselves.

1.6 Abstracts of the Chapters

A preview of the remaining chapters is given below in the form of an abstract for each.

Chapter 2: Semi-Geostrophic Models

This chapter discusses the nature of geostrophic momentum approximation, which is the basis of an SG model capable of simulating fronts. The governing equations of a 2D SG model are then formulated, leading to a derivation of the SG tendency equation and SE equation. We examine a canonical solution of the SE equation that reveals the dependence of a frontal transverse circulation upon the three key parameters. We go over the algorithm of a complete analysis of an SG model in physical coordinates. For a 3D SG model, we derive and solve two coupled diagnostic equations for the transverse circulation. They are a generalization of the SE equation.

Chapter 3: Construction of Two Initial States for Modeling Fronts

This chapter constructs the initial disturbances to be used in the subsequent simulations of surface fronts in two generic 2D models. One version of the model has the Eady flow as its reference state. We go over a complete modal instability analysis of this basic flow, present all of its instability properties and discuss the essence of this instability. The other version of the model has a modified but equivalent basic state of the Charney model (e-Charney model) as a reference state. A corresponding instability analysis is presented. These materials are pedagogically interesting in their own right.

Chapter 4: Two-Dimensional Semi-Geostrophic Frontal Dynamics

This chapter discusses the basic dynamics in the life-cycle of idealized 2D fronts in two illustrative analyses using the initial states elaborated in Chapter 3. The results of the 2D SG Eady frontal model analysis illustrates how and why fronts develop on the top and bottom surfaces of a minimal setting. In addition, the impact of dissipation is discussed. We next illustrate that only a surface front would develop in a 2D SG e-Charney frontal model. We finally discuss the intrinsic limitations of 2D frontal models.

Chapter 5: Three-Dimensional Semi-Geostrophic Frontal Dynamics

Three-dimensional frontal dynamics are discussed in the context of a 3D SG model analysis with a simplified baroclinic jet as its reference state. The illustrative analysis is formulated and performed in physical coordinates instead of geostrophic coordinates. Broadly realistic cold fronts and warm fronts are shown to develop simultaneously in this model. It is found that the differential thermal advection by the geostrophic wind and ageostrophic wind components in the cold and warm frontal regions contributes to the distinct difference in the intensity of those fronts. We also present a complete analysis of the energetics of the disturbance in this SG model.

Chapter 6: Impact of Latent Heat Release on Fronts

Several major parameterization schemes for condensational heating are first reviewed. We apply one of them in an illustrative analysis with a 2D SG frontal model. With relevant values of the parameters, it is found that the model surface front with such heating is substantially stronger and narrower than the counterpart dry model front. The frontal collapse also occurs sooner. The sensitivity to the heating profile is examined. Qualitatively similar impacts of the self-induced condensational heating are found in an application to a counterpart 3D SG moist model.

Chapter 7: General Frontal Dynamics

This is a long chapter that addresses the nature of general frontal dynamics as manifested in a frontal model that has virtually no dynamical constraints. The discussions are based on two investigations using a high-resolution **A**dvanced **R**esearch **W**eather **R**esearch **F**orecasting model that is run in the non-hydrostatic mode including diabatic heating. The basic state is a realistic baroclinic jet straddling a dynamic tropopause. Quite realistic surface cold and warm fronts, as well as a ULF, are simulated at the same time. They are shown to be integral parts of an intensifying synoptic baroclinic wave. We particularly investigate why cold fronts are characteristically much narrower, longer, and more intense than warm fronts. We particularly delineate the structural and dynamic properties of the model ULF. We show that while the appropriate metric of surface frontogenesis is $\frac{D}{Dt}|\nabla_2\theta|$, the metric for upper-level frontogenesis is $\frac{D}{Dt}|\nabla\theta|$, where ∇_2 and ∇ are 2D and 3D gradient operators respectively. We finally investigate whether, how, and why a ULF could synergistically interact with a surface cold front.

Chapter 8: Concluding Remarks

Cyclogenesis gives rise to frontogenesis, not vice versa. Specifically, fronts emerge from an intensifying extratropical cyclone and are sites of severe weather elements. Fronts are a product of *scale-interaction* involving planetary-, synoptic-, and meso-scale components of the flow even in a dry model setting. The meso- and small-scale components in a moist model front have additional important impacts of condensational heating. Fronts play the role of an essential link in the global cascade of energy. The ubiquitous cyclone–front–convection systems in the atmosphere have a high degree of disorderliness and therefore lead to maximization of its total entropy.

2
Semi-Geostrophic Models

2.1 Introductory Remarks

In this chapter we show the formulation of a class of simplest possible and yet adequate models for the purpose of delineating the fundamental aspect of frontal dynamics. Applications for several versions of this model will be presented and applied in Chapters 4 and 5, in conjunction with suitable initial conditions to be constructed in Chapter 3. We will wait until Chapter 7 to take on the more ambitious task of simulating atmospheric fronts in a realistic setting with a model that is much less dynamically constrained.

Let us begin by taking account of some broad conceptual considerations. The fact that fronts are observed in the oceans as well as in the atmosphere plainly suggests that condensational heating is not indispensable as far as frontal formation is concerned, although precipitation is always prevalent in frontal regions. Thus, it would suffice for a basic model of frontal dynamics to be dry. Furthermore, the observational evidence suggests that internal gravity waves do not seem to play an essential role during frontal development, even though they are likely to be excited toward the end of a life-cycle of fronts when the local shear becomes very strong and the stratification very weak. Thus, even a dry primitive equation model, in which gravity waves are present, would be more general than necessary for investigating the intrinsic dynamics of fronts. Also, although friction near the ground is expected to be important in limiting the width and intensity of fronts, it is believed to have only quantitative but not qualitative impact on the development of fronts. In light of these considerations, it would be reasonable to expect a *dry inviscid balance model* capable of capturing the frontal dynamics.

The simplest balance model is a quasi-geostrophic (QG) model for large-scale atmospheric flow. As mentioned in Chapter 1, bona fide fronts cannot be simulated in a QG model framework. What we should be looking for is a model framework that is neither too general nor too restrictive: a Goldilocks sort of model.[1]

[1] The children's story "Goldilocks and the Three Bears" by Robert Southey (1837).

The critical process missing in a QG model is the advection of heat and momentum by the ageostrophic velocity component, even though the ageostrophic velocity component is an order of magnitude weaker than the geostrophic velocity component in a synoptic disturbance. That process can be incorporated with the use of an approximation known as the *geostrophic momentum approximation*. A model invoking this approximation is known as a *semi-geostrophic (SG) model*. The focus of this chapter is to discuss how one would formulate an SG model in a 2D or a 3D setting.

We will take the following approach in a first investigation of the dynamical nature of atmospheric fronts:

(1) to formulate the simplest possible mathematical model that would adequately depict a phenomenon under consideration;
(2) to work out the methodology of analyzing the governing equations of such a model;
(3) to construct an appropriate state of the model at the initial time;
(4) to determine how this state will change over time by applying the said methodology of analysis; and
(5) to interpret the model result on the basis of the assumptions and fundamental concepts underlying the model.

We will specifically establish the geostrophic momentum approximation and identify the condition of its validity in Section 2.2. It will then be applied to an inviscid adiabatic rotating Boussinesq fluid model on a f-plane in a cyclical channel domain. The beta-effect will later be included through the use of an additional parameter. We begin by going over, in Section 2.3, the formulation of a 2D SG model, some of its intrinsic properties as well as the method of analysis. The formulation will be extended to the general case of a 3D SG model in Section 2.4. The method of analysis for a 3D model is considerably more involved and is elaborated in Section 2.5. These discussions would serve as background information for performing the particular illustrative analyses in Chapters 4 and 5.

2.2 Geostrophic Momentum Approximation

The ageostrophic component influences the evolution of the geostrophic component of a flow through different processes. In the QG theory, the ageostrophic flow component exerts influence via the processes of vortex stretching associated with the Coriolis parameter and adiabatic heating/cooling associated with basic stratification. Although the ageostrophic velocity component could also advect momentum and heat, that process is not represented in the QG theory. Eliassen (1948) introduces an approximation that replaces acceleration of the horizontal velocity by acceleration of the geostrophic component of the flow, but keeps the total derivative operator intact, namely:

$$\frac{DV}{Dt} \approx \frac{DV_g}{Dt}$$
$$\frac{D}{Dt} = \frac{\partial}{\partial t} + u_{total}\frac{\partial}{\partial x} + v_{total}\frac{\partial}{\partial y} + w_{total}\frac{\partial}{\partial z}. \tag{2.1}$$

This is the geostrophic momentum approximation, which is the cornerstone of an SG model.

This approximation is compatible with several observational evidences. One is that the ageostrophic velocity component is largely normal to a front, in contrast to the geostrophic velocity component in a frontal region, which is closely parallel to a front. Furthermore, the geostrophic velocity is considerably stronger than the ageostrophic velocity (usually by a factor of five or more) and the width of a front is much shorter than its length (by more than a factor of ten). It follows that the horizontal advection of momentum by both the geostrophic and ageostrophic components can be comparable during frontal development. The advection of momentum and heat by the vertical velocity (which is part of the 3D ageostrophic circulation) is also comparably large wherever the local vertical gradient of potential temperature and geostrophic velocity are large.

Let us now go over a systematic deduction of how the momentum equation for a rotating fluid without friction might be simplified. *In the interest of clarity, we for the moment use the subscripts "g" and "a" to refer to the geostrophic and ageostrophic components respectively, although we will indicate them with superscripts "∧" and " ∼ " in the rest of the book unless specified otherwise, i.e.,* $\boldsymbol{V_g} = (u_g, v_g) \equiv (\hat{u}, \hat{v})$, $\boldsymbol{V_a} = (u_a, v_a) \equiv (\tilde{u}, \tilde{v})$, $\boldsymbol{V} = \boldsymbol{V_g} + \boldsymbol{V_a}$. The horizontal momentum equation can be written as

$$\begin{aligned}
\mathbf{V} &= \boldsymbol{V_g} + \frac{1}{f}\boldsymbol{k}\times\frac{D\boldsymbol{V}}{Dt} \\
&= \boldsymbol{V_g} + \frac{1}{f}\boldsymbol{k}\times\frac{D}{Dt}\left(\boldsymbol{V_g} + \frac{1}{f}\boldsymbol{k}\times\frac{D\boldsymbol{V}}{Dt}\right) \\
&= \boldsymbol{V_g} + \frac{1}{f}\boldsymbol{k}\times\frac{D\boldsymbol{V_g}}{Dt} - \frac{1}{f^2}\frac{D^2\boldsymbol{V}}{Dt^2} \\
&\approx \boldsymbol{V_g} + \frac{1}{f}\boldsymbol{k}\times\frac{D\boldsymbol{V_g}}{Dt}.
\end{aligned}$$

Neglecting the term $\frac{1}{f^2}\frac{D^2V}{Dt^2}$ in the last step is justifiable if the time scale for change in velocity following a fluid particle is considerably longer than $f^{-1} \approx 3$ h. In other words, the Lagrangian Rossby number should be small, $Ro_L = \frac{|DV_g/Dt|}{|fk \times V_g|} \ll 1$. This is less restrictive than requiring a small Eulerian Rossby number, $Ro_e = \frac{U}{fL} \ll 1$, where U and L are the horizontal velocity and length scales of the flow under consideration. Ro_L depends only on the component of $\frac{D\mathbf{V}_g}{Dt}$ being normal to $\mathbf{V}_g$. Therefore, it requires the rate of change of horizontal wind direction to be much less than f, but does not restrict the rate of change of its magnitude. It means that a trajectory should not change its direction by more than an angle of $\pi/4$ in 24 hours. Typical trajectories from a weather forecasting model indicate that this requirement is generally satisfied in a large-scale flow (see figure 2.1 in Cullen 2006).

Hence, the simplified horizontal momentum equation for the fluid under such condition is justifiably written as:

$$\frac{D\boldsymbol{V}_g}{Dt} = -f\boldsymbol{k} \times \boldsymbol{V}_a. \tag{2.2}$$

Furthermore, for the disturbances to be considered in an application of the geostrophic momentum approximation, the time scale is still long compared to the development time of convective clouds. Thus, hydrostatic balance would be a justifiable approximation for the vertical momentum equation

There is a subtle difference between the line of reasoning for the simplification above from the reasoning behind a derivation of the QG theory. The most systematic derivation of the latter is done with a standard asymptotic expansion in terms of the Eulerian Rossby number (Pedlosky, 1987). That technique is used to identify all terms of the same order of magnitude in each governing equation and thereby arrive at a self-consistent approximate version of it. For example, let us see what one would conclude when we seek to estimate the order of magnitude of the advective acceleration. The expansion would yield

$$\boldsymbol{V} \cdot \nabla \boldsymbol{V} = \boldsymbol{V}_g \cdot \nabla \boldsymbol{V}_g + \boldsymbol{V}_g \cdot \nabla \boldsymbol{V}_a + \boldsymbol{V}_a \cdot \nabla \boldsymbol{V}_g + \boldsymbol{V}_a \cdot \nabla \boldsymbol{V}_a.$$

For a mid-latitude cyclone type of disturbance, it is definitely true that $V_g \gg V_a$. In the usual asymptotic expansion analysis, it is assumed to have one single horizontal length scale. It would follow that the first term $V_g \cdot \nabla V_g$ would seem to be the sole dominant term and the others would be an order of magnitude smaller. But the last argument is not valid in the neighborhood of a frontal zone when a front is evolving toward its mature stage. In that stage, the local spatial scale of temperature and velocity is substantially shorter than the scale of the disturbance as a whole. In particular, $\boldsymbol{V}_a$ is directed much more normal to the front itself than $\boldsymbol{V}_g$. It follows that the products $\boldsymbol{V}_g \cdot \nabla \boldsymbol{V}_g$ and $\boldsymbol{V}_a \cdot \nabla \boldsymbol{V}_g$ could be comparable. Therefore, both terms need to be retained in order to adequately capture the transient dynamic of the frontal evolution. We should introduce similar simplification to the thermal advection in the thermodynamic equation as well. That is the crucial reasoning behind the formulation of an SG model.

It is clear that an SG model would be reduced to a QG model if we additionally approximate the total derivative as $\frac{D}{Dt} \approx \frac{D_g}{Dt} \equiv \frac{\partial}{\partial t} + \boldsymbol{V}_g \cdot \nabla$ in (2.2). Also, in the context of linear dynamics prior to frontal formation, the advection of momentum and heat by the ageostrophic velocity component would be negligible compared to that by the geostrophic velocity component. It follows that an elementary solution of an SG model would be just a Rossby wave as in a QG model.

2.3 Two-Dimensional SG Frontal Model

Frontal development is a complicated time-dependent nonlinear dynamical process. It would be prudent to take the first stab at the problem with the simplest possible version of an SG model. That would be a 2D adiabatic inviscid SG model. Since fronts do not form in isolation but develop within a large-scale disturbance under the influence of a certain background flow, we introduce the latter in a model with a specific objective in mind in the context of the following definition:

$$\textit{total state} = (\textit{reference state}) + (\textit{time dependent state}).$$

For simplicity, we first consider the case of a 2D disturbance developing under the influence of a certain background state that has a basic stratification and a simple zonal shear flow. All properties of such a disturbance do not vary in the direction of the front, say $\frac{\partial}{\partial y} = 0$ by definition. The reference state would be characterized by $\overline{u}(z), \overline{\theta}(y,z), \theta_o(z)$ where $\overline{u}$ is a purely baroclinic shear flow in geostrophic balance with a certain $\overline{\phi}(y,z)$ and $\overline{\theta}$ is related to $\overline{u}$ by the thermal wind equation, $\overline{\theta}_y = -\frac{\theta_{oo} f}{g} \overline{u}_z$. f is the Coriolis parameter; g is gravity; and θ_{oo} is a constant. For simplicity, we further consider $\frac{d\theta_o}{dz} \gg \frac{\partial \overline{\theta}}{\partial z}$. Subscripts here generally stand for partial derivatives. The background stratification is expressed in terms of a Brunt–Vasaila frequency N such that $N^2 = \frac{g}{\theta_{oo}} \frac{d\theta_o}{dz}$, which is either a constant or a function of z. The variables of the departure state consist of a geostrophic velocity component $(0, \hat{v}, 0)$ and a potential temperature field $\hat{\theta}$ in thermal wind balance $\hat{\theta}_x = \frac{\theta_{oo} f}{g} \hat{v}_z$. Note that since this model domain is laterally unbounded, there would be no constraint to the total v field. The departure state also has an ageostrophic velocity component $(\tilde{u}, 0, \tilde{w})$ on the (x, z) cross-section. It is worth emphasizing that while the reference state $\overline{u}(z), \overline{\phi}(y,z), \overline{\theta}(y,z), \theta_o(z)$ introduced at the initial time could serve as a forcing for the subsequent development of the departure field $\hat{v}, \hat{\theta}$ and $(\tilde{u}, 0, \tilde{w})$, the total state at subsequent time could depart substantially from the reference state.

The governing equations of this model are then

$$\begin{aligned}
&\frac{D\hat{v}}{Dt} &&= -f\tilde{u} \\
&\frac{D(\overline{\theta}+\hat{\theta})}{Dt} + \hat{v}\overline{\theta}_y + \tilde{w}\frac{d\theta_o}{dz} &&= 0 \\
&\tilde{u}_x + \tilde{w}_z &&= 0 \\
&\hat{\theta}_x &&= \frac{\theta_{oo}f}{g}\hat{v}_z
\end{aligned}, \tag{2.3a–d}$$

where $\frac{D}{Dt} = \frac{\partial}{\partial t} + (\overline{u}+\tilde{u})\frac{\partial}{\partial x} + \tilde{w}\frac{\partial}{\partial z}$ is the total derivative. It should be emphasized that the momentum equation in the x-direction is simply the geostrophic balance, $0 = -\frac{1}{\rho_{oo}}\hat{p}_x + f\hat{v}$, and the vertical momentum equation is simply the hydrostatic balance, $0 = -\frac{1}{\rho_{oo}}\hat{p}_z + \frac{g\hat{\theta}}{\theta_{oo}}$. They are embodied in the thermal wind relation (2.3d). The domain is $-\infty > x < \infty$, $0 \leq z \leq D$. The boundaries $z = 0$, D are rigid surfaces through which there is no mass flux. The problem is to determine the solution that describes the evolution of the departure flow field $(\tilde{u}, \hat{v}, \tilde{w})$ and departure temperature $\hat{\theta}$ on the cross-section (x, z) starting from a certain initial state for a particular reference state $\overline{u}(z), \overline{\theta}(y, z), \theta_o(z)$. Both the geostrophic and ageostrophic flow components could play active roles in the development of frontogenesis.

The first successful study of frontal dynamics was done in the context of this 2D model using the Eady flow as a reference state, namely $\overline{u} = \lambda z$, $\overline{\theta} = -\frac{f\lambda\theta_{oo}}{g}y$ and a stable stratification N^2 (Hoskins and Bretherton, 1972). For the case of using an unstable normal mode as an initial disturbance, it is possible to obtain an analytic solution only by judiciously making use of the so-called geostrophic coordinates defined as $(X = x + f^{-1}\hat{v},\ Z = z)$. Such coordinate transformation renders the governing equations as if they were linear, albeit the physical meaning of the corresponding transformed dependent variables is less transparent. The solution transformed back to physical space reveals the formation of a front at each model boundary intensifying without bound (toward a singularity in finite time).

We do not have to use geostrophic coordinates when we seek a numerical solution. Such a solution in physical coordinates can be straightforwardly determined. The advantage of solving the nonlinear problem in physical coordinates with a numerical method is that it would allow us to readily perform frontal simulation using any reference state and any initial condition for different purposes.

It is noteworthy that the system of (2.3) cannot be numerically solved by simply coding them as-is. Should we separately calculate the changes of $\hat{v}$ and $\hat{\theta}$ directly with (2.3a,b) for one time-step, chances are that their new values would not be mutually compatible because it is unlikely that they would exactly satisfy the thermal wind relation (2.3d). Furthermore, we could not evaluate the $(\tilde{u}, 0, \tilde{w})$ at the new time-step with this form of the equations. Therefore, it is necessary to recast the equations to a more amenable form.

2.3.1 Two-Dimensional SG Potential Vorticity and Sawyer–Eliassen Equation

We may define an SG potential vorticity for this model as

$$\begin{aligned} q &= -\hat{v}_z\hat{\theta}_x + (f+\hat{v}_x)(\theta_{oz}+\hat{\theta}_z) \\ &= \frac{f\theta_{oo}}{g}\left(-\psi_{xz}^2 + (f+\psi_{xx})\left(\frac{N^2}{f}+\psi_{zz}\right)\right). \end{aligned} \tag{2.4}$$

It is a function of the geostrophic streamfunction ψ alone. This is an important property because it is a materially invariant property of this flow. Let us first verify this to be true. On the basis of the expression (2.4) we have

$$\begin{aligned} \frac{g}{f\theta_{oo}}\frac{Dq}{Dt} &= -2\hat{v}_z(\hat{v}_{zt} + (\overline{u}+\tilde{u})\hat{v}_{zx} + \tilde{w}\hat{v}_{zz}) \\ &\quad + (f+\hat{v}_x)(\psi_{zzt} + (\overline{u}+\tilde{u})\psi_{zzx} + \tilde{w}\psi_{zzz}) \\ &\quad + \left(\frac{N^2}{f}+\psi_{zz}\right)(\hat{v}_{xt} + (\overline{u}+\tilde{u})\hat{v}_{xx} + \tilde{w}\hat{v}_{xz}). \end{aligned} \tag{2.5}$$

This can be rewritten by making use of (2.3a,b,c) as

$$\frac{g}{f\theta_{oo}}\frac{Dq}{Dt} = \hat{v}_z\left[(f+\hat{v}_x)\tilde{u}_z + 2\hat{v}_z\tilde{w}_z - \left(\frac{N^2}{f}+\psi_{zz}\right)\tilde{w}_x + 2\overline{u}_z\hat{v}_x\right]. \tag{2.6}$$

On the other hand, we may introduce an ageostrophic potential χ such that $\tilde{u} = -\chi_z$ and $\tilde{w} = \chi_x$ in light of (2.3c). Upon eliminating the time derivative terms between (2.3a,b) with the use of the thermal wind balance, we also obtain a diagnostic equation, namely:

$$f(f+\hat{v}_x)\chi_{zz} - \frac{2g\hat{\theta}_x}{\theta_{oo}}\chi_{xz} + \left(N^2 + \frac{g\hat{\theta}_z}{\theta_{oo}}\right)\chi_{xx} = 2f\,\overline{u}_z\hat{v}_x. \tag{2.7}$$

We consider a domain with rigid boundaries at the top and bottom. The boundary conditions are then $\chi(x,0,t) = \chi(x,D,t) = 0$. Equation (2.7) is known as the *Sawyer–Eliassen (SE) equation* (Sawyer, 1956; Eliassen, 1962). It is pertinent to note that the square bracketed quantity on the RHS of (2.6) can be rewritten as

$$(f+\psi_{xx})\chi_{zz} - 2\psi_{xz}\chi_{xz} + \left(\frac{N^2}{f}+\psi_{zz}\right) = 2\overline{u}_z\psi_{xx}. \tag{2.8}$$

Except for a constant factor f in every term, this is identical to the SE equation, (2.7). Thus, we may conclude that q defined by (2.4) is indeed a materially conservative property of a 2D SG model. Equation (2.7) is also a very useful equation with which we determine χ whenever the geostrophic velocity field is known, thereby enabling us to determine a self-compatible ageostrophic velocity field.

2.3.2 Properties of the Sawyer–Eliassen Equation

The SE equation, (2.7), is a second-order partial differential equation that has variable coefficients on the LHS and an inhomogeneous term on the RHS. They are functions of the geostrophic velocity $\hat{v}(x, z, t)$ and related potential temperature $\hat{\theta}(x, z, t)$ apart from the parameters of the reference state. The appropriate boundary conditions would be $\tilde{u} = -\chi_z = 0$ at the vertical boundaries and $\tilde{w} = \chi_x = 0$ at the top and horizontal boundaries. It amounts to imposing $\chi = 0$ at all boundaries of the domain. Equation (2.8) has a generic form $A\chi_{zz} + B\chi_{xz} + C\chi_{xx} = F$ where $A = f(f + \hat{v}_x)$ is referred to as the *inertial stability parameter*, $B = \frac{-2g\hat{\theta}_x}{\theta_{oo}}$ as the *baroclinicity parameter* and $C = N^2 + \frac{g\hat{\theta}_z}{\theta_{oo}}$ as the *local static stability parameter*. In the case of an Eady flow $\bar{u} = \lambda z$ as the reference state, we have $F = 2f\lambda\hat{v}_x$. Whether or not there exists a solution of χ hinges upon whether or not it is an elliptic differential equation. Existence of a solution requires the following condition of inequality:

$$B^2 - 4AC < 0 \Rightarrow \left(\frac{g\hat{\theta}_x}{\theta_{oo}}\right)^2 - [f(f + \hat{v}_x)]\left[N^2 + \frac{g\hat{\theta}_z}{\theta_{oo}}\right] < 0. \tag{2.9}$$

Two factors in this setting have a *stabilizing* influence on a flow: stable stratification, $(\theta_{total})_z > 0$, and rotation of the reference frame, f^2. Two other factors have a *destabilizing* influence: vertical shear and cross-front shear of the along-front velocity, $\hat{v}_z$ and $\hat{v}_x$. The existence of solution for χ in this model presupposes that the evolving along-front velocity remains a balanced flow. It means that a cross-front circulation would exist only if the net effect of the two stabilizing factors is stronger than that of the two destabilizing factors. To summarize, if we have

(1) a statically stable stratification, $C = N^2 + \frac{g\hat{\theta}_z}{\theta_{oo}} > 0$,
(2) an inertially stable along-front velocity, $A = f(f + \hat{v}_x) > 0$, and
(3) a sufficiently weak vertical shear in the along-front velocity,

we would have

$$(f\hat{v}_z)^2 < [f(f + \hat{v}_x)]\left[N^2 + \frac{g\hat{\theta}_z}{\theta_{oo}}\right]. \tag{2.10}$$

Then, under this condition, there would be a self-compatible ageostrophic circulation on the cross-front plane. Also note that

$$\begin{aligned}
&[f(f+\hat{v}_x)]\left[N^2+\frac{g\hat{\theta}_z}{\theta_{oo}}\right]-(f\hat{v}_z)^2>0\\
&\rightarrow (f+\hat{v}_x)\left(\frac{\partial\theta_{total}}{\partial z}\right)-\hat{v}_z\frac{\partial\hat{\theta}}{\partial x}>0\\
&\rightarrow \varsigma_{absolute}\cdot\nabla\theta_{total}>0.
\end{aligned} \tag{2.11}$$

The condition for the existence of a solution of the SE equation then amounts to requiring the potential vorticity to have positive values everywhere in the domain. Whenever this condition breaks down, the geostrophic flow would become dynamically unstable (Mak, 2011). There would no longer be a self-compatible ageostrophic circulation in the system.

The ageostrophic potential in QG theory is governed by the commonly known "omega equation," namely:

$$f^2\chi_{zz}+N^2\chi_{xx}=2f\bar{u}_z\psi_{xx}.$$

This second-order partial differential equation is always elliptic as long as the reference state has a stable stratification ($N^2 > 0$). The disturbance geostrophic velocity component only appears on the RHS. In other words, potential vorticity in SG theory plays the same dynamical role as static stability in quasi-geostrophic theory from the perspective of ageostrophic circulation. The absence of the geostrophic variables on the LHS in the omega equation means that there is no direct feedback influence of the ageostrophic flow on QG frontal development.

We can use the SE equation not only to compute the ageostrophic circulation for a given geostrophic circulation, but also to explain why there can be inexorable positive feedback leading to frontal formation. The ageostrophic circulation in response to a given geostrophic circulation is analogous to the response of any physical system subject to an external forcing. The strength of the response would depend on the structural properties of the system under consideration, as well as the intensity of a forcing. For a mechanical system like a rod or spring, the relevant property is its "stiffness." For a DC (direct current) electric circuit, the relevant property is "resistance"; for an AC (alternating current) circuit, it is "impedance." In the case of an SG fluid system, the pertinent structural property is expressed by a combination of three parameters ($B^2 - AC$), where A is the *inertial stability parameter*, *B the baroclinicity parameter*, and C the *local static stability parameter*. This property is equivalent to the stiffness/resistance/impedance of the other systems. The three parameters of an SG model depend on the geostrophic circulation itself. So as the geostrophic circulation changes, the equivalent "stiffness" of the system and hence the strength of the response to a unit of geostrophic forcing would change. Specifically, as the geostrophic circulation becomes stronger, the stiffness would become weaker, resulting in a stronger ageostrophic response . It would in turn lead to an even stronger geostrophic circulation as one would find out when one solves the governing equation for the geostrophic circulation itself. Here lies the nonlinear feedback responsible for an inexorable frontal development .

2.3.3 Canonical Solutions of the Sawyer–Eliassen Equation

In general, it is only feasible to numerically solve the SE equation at a particular instant because the parameters A, B, and C typically are functions of the space variables as well as time. It would be instructive to learn about the characteristics of the solution of the SE equation in the simplest possible situation. Suppose all its coefficients *happen to be constants* satisfying the inequality $B^2 - 4AC < 0$, the SE equation can be rewritten in a form known as Poisson equation by transforming the physical coordinates (x, z) to geostrophic coordinates (X, Z) where $X = x + f^{-1}\hat{v}$ and $Z = z$. It follows from this definition that we have $dX = dx(1 + f^{-1}\hat{v}_x)$. At a location where the geostrophic relative vorticity is cyclonic $\hat{v}_x > 0$, such as across a front, the *X-coordinate* would be locally stretched since $\frac{dX}{dx} > 1$. In other words, the locally short-length scale across a front in physical space would appear much longer in the geostrophic space. Furthermore, by introducing appropriate scaling the Poisson equation would take on a canonical form $\frac{\partial \chi^*}{\partial X^{*2}} + \frac{\partial \chi^*}{\partial Z^{*2}} = F^*$, where the superscript * refers to nondimensional quantities. The solution is subject to the boundary condition $\chi = 0$. One may express the solution in an integral form. The integrand would be the product of a two-dimensional Green's function and the forcing function. The Green's function is the response to a forcing equal to unit delta function. It plays the role of a building block in the construction of the response to any general forcing. The Green's function can be made to satisfy the homogeneous boundary condition with the method of images. The solution $\chi^*(X, Z)$ may be finally transformed back to physical space, $\chi(x, z)$. This methodology entails considerable algebraic works. The analysis can be somewhat simplified by first applying Fourier transform for the X-dependence of $\chi^*$and F^*. The Z-dependence of each Fourier component of χ^*can then be expressed in an integral form in terms of a one-dimensional Green's function and the corresponding spectral component of F^*. The boundary condition is to be satisfied by each spectral Green's function. Hakim and Keyser (2001) report such a canonical solution for four cases relative to a control case, as listed in Table 2.1.

Table 2.1 lists some properties of the ageostrophic streamfunction in response to a unit forcing for four cases: (1) large inertial stability, $A = f(f + \hat{v}_x)$, (2) large baroclinicity, $B = \frac{-2g\hat{\theta}_x}{\theta_{oo}}$, (3) large static stability, $C = N^2 + \frac{g\hat{\theta}_z}{\theta_{oo}}$, and (4) a forcing located near a boundary. The solution for the case of large inertial stability is shown in Figure 2.1a. It reveals that the major axis of the ellipse (response) is oriented mostly

Table 2.1. Parameter values for SG control and test cases

Case	F^2	S^2	N^2	qSG	L	R	ψ_0	W_0	ua_0
Control	1	3	1	0.91	1000	0.734	9.55	1.00	0.955
Large F^2	2	3	1	1.91	707	0.806	4.77	0.69	0.477
Large S^2	1	6	1	0.64	1000	0.500	9.55	1.19	0.955
Large N^2	1	3	2	1.91	1414	0.806	9.55	0.69	0.955

Source: taken from Hakim and Keyser (2001).

in the direction of absolute momentum surfaces (dashed line). Figure 2.1c reveals that the major axis of the response for the case of large local static stability is oriented mostly in the direction of the isentropic surfaces, $\theta_o + \theta'$. Figure 2.1b reveals that the major axis of the response for the case of large baroclinicity is oriented in a direction in between that of the absolute momentum surface and isentropic surface. Figure 2.1d reveals that the presence of a rigid boundary strongly skews the ellipse due to the strong constraint of the flow at a rigid surface. One can readily deduce the corresponding distribution of the ageostrophic velocity components from the ageostrophic streamfunction in each case.

We will further elaborate in Section 7.6.2 on another canonical solution for a case of having two-point forcing judiciously placed at two locations in the domain. The resulting response is particularly relevant to the consideration of possible interaction between an upper-level front and a surface front.

2.3.4 Two-Dimensional SG Tendency Equation and Algorithm

We now proceed to discuss how one might obtain a complete solution under a general condition in the context of a 2D SG model. That would portray the formation and evolution of a frontal circulation. Let us first consider the technical issue of getting a solution from the perspective of potential vorticity. The material invariance of potential vorticity is in essence a prognostic equation. For the reference state under consideration, it is

$$q_t = -(\overline{u} + \tilde{u})q_x - \tilde{w}q_z. \tag{2.12}$$

It is instructive to note before we proceed that if we were to drop the terms representing the advection of heat and momentum by the ageostrophic velocity component in a 2D SG model (amounts to reducing it to a QG model), we would have $\frac{D}{Dt} = \frac{\partial}{\partial t} + \overline{u}\frac{\partial}{\partial x}$ and (2.3) would be reduced to a set of linear equations. Since we can categorically state that there could be no scale contraction in a linear system, it follows that frontal formation would not be possible. Although the geostrophic flow component of an unstable 2D baroclinic wave in the course of its intensification would tend to increase the temperature gradient of the wave field by thermal advection, it is largely counteracted by the adiabatic heating/cooling associated with the

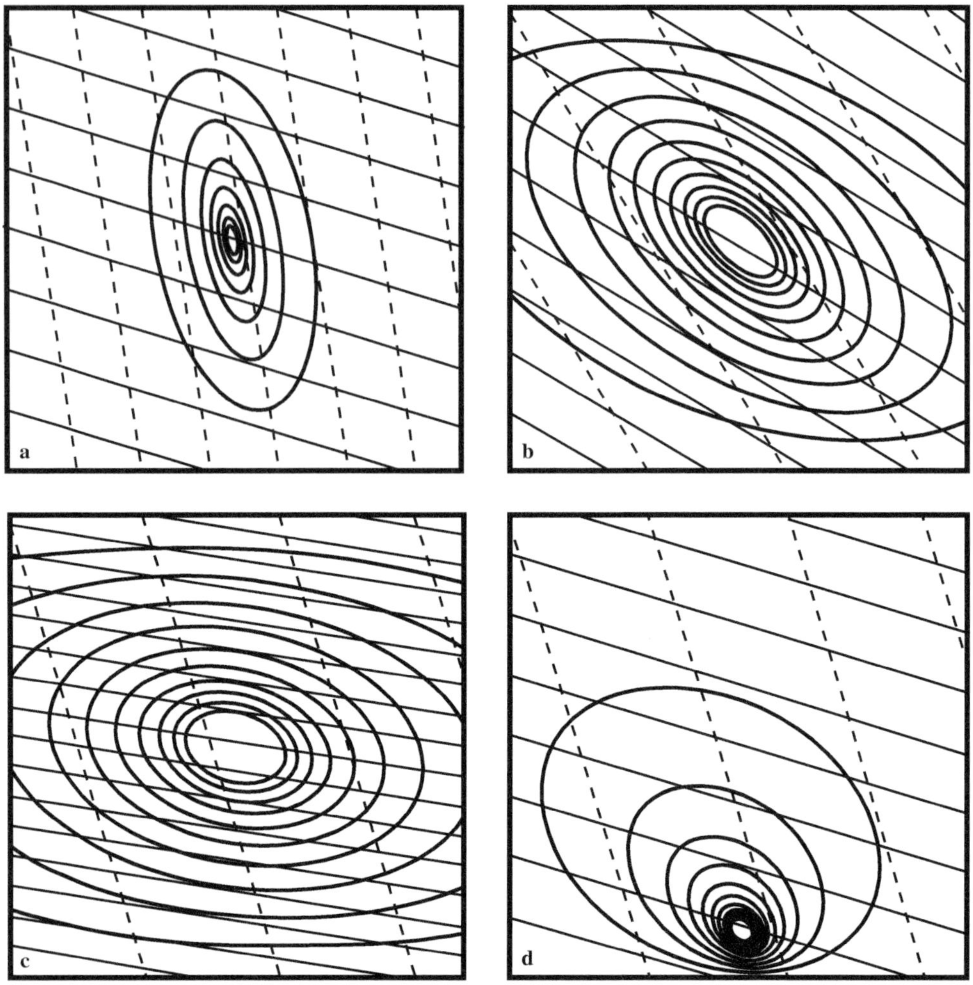

Figure 2.1. Semi-geostrophic solutions for (a) large inertial stability, (b) large baroclinicity, (c) large static stability, and (d) near-boundary point source. The streamfunction is given by thick lines every 2×10^3 m^2 s^{-1}, potential temperature is given by solid lines every 4 K, and absolute momentum is given by thin dashed lines every 30 m s^{-1}. Parameter values are given in Table 2.1; the values for the control case apply in (d) (taken from Hakim and Keyser, 2001).

vertical motion. The geostrophic process would lead to a contraction of scale and hence frontogenesis only if the initial disturbance consists of multiple unstable waves. Such a process is sometimes referred to as *quasi-geostrophic frontogenesis*. On the other hand, the enhancement of thermal gradient in an SG dynamical framework due to the thermal advection by the ageostrophic circulation would proceed unopposed. That process would lead to inexorable development of a front. It is sometimes referred to as *ageostrophic frontogenesis*. Thus, this especially simple model enables us to conclude that the advection of momentum and heat by the ageostrosphic velocity plays an indispensable role in frontal dynamics. In other words, the SG model is indeed the simplest possible model we need to use in order to capture the intrinsic dynamics of fronts. It would be straightforward to numerically integrate (2.12) to get the change of the potential vorticity field q over one time-step when the state of the disturbance is known at any time. However, it would not be easy to deduce the corresponding change in the geostrophic streamfunction field ψ by solving (2.4) because the latter is a nonlinear partial differential equation.

It is useful to derive an alternative prognostic equation by suitably combining (2.3a,b). For the moment, let us assume that the basic stratification N does not vary in space. Taking the derivative of (2.3a) with respect to x and (2.3b) with respect to z, we add the resulting two equations. With the use of the hydrostatic balance, we would then obtain the following equation, which may be referred to as a *2D SG tendency equation*, namely:

$$P_t = -(\overline{u}+\tilde{u})P_x - \tilde{w}P_z - \tilde{u}_x\psi_{xx} - \tilde{w}_z\frac{f^2}{N^2}\psi_{zz} - \left(\tilde{w}_x + \tilde{u}_z\frac{f^2}{N^2}\right)\psi_{xz}, \tag{2.13}$$

$$P = \psi_{xx} + \frac{f^2}{N^2}\psi_{zz}. \tag{2.14}$$

With the condition $\tilde{w}=0$ at $z = 0, D$, (2.3b) is reduced to

$$@\ z=0, z_{top}\ \psi_{zt} = -(\overline{u}+\tilde{u})\psi_{xz} - \psi_x\overline{\theta}_y. \tag{2.15}$$

P is identifiable with the QG potential vorticity. This alternative approach was first noted by Schubert (1985), although he did not apply it in an analysis.

We may perform the following algorithm for an analysis of 2D frontal dynamics. With a particular initial state, we integrate (2.13) with (2.15) to determine the change of the P field over one time-step with a standard predictor-corrector numerical scheme. Next we obtain the corresponding change in the ψ field by solving the linear partial differential equation (2.14). The next step is to determine the corresponding new ageostrophic velocity field by solving (2.7) for χ and thereby those of $\tilde{u}$ and $\tilde{w}$ fields at the new time-step as well. By now we can resume marching forward in time by repeating this sequence of steps above as long as inequality (2.11) is met. However, should that inequality be violated at certain times when a model front has become very intense, the numerical integration would stop because all fields would soon be overwhelmed by numerical noise. Such a model state may be interpreted as frontal collapse. We would have reached the limit of validity of this SG model at that point.

The first illustrative analysis to be presented in Section 4.2 is an application of the algorithm described above for the case of a zonal baroclinic flow with a constant shear in the reference state. A suitable initial disturbance for that frontal simulation is constructed on the basis of a corresponding instability analysis to be presented in Section 3.2. That is the simplest setting that would generate what one might call generic fronts. We will generalize the model in two distinctly different ways.

We may, for instance, include the beta-effect in an SG model through the use of an additional parameter β. The RHS of (2.13) would have an extra term, namely $-\beta\psi_x$. The reference state will also be slightly modified for that simulation so that a model front would only develop at the surface. A suitable initial disturbance for that frontal simulation is again constructed on the basis of a corresponding instability analysis in Section 3.3. We will demonstrate frontal formation in each case.

2.4 Three-Dimensional SG Frontal Model

We will see that the two-dimensionality of an SG model gives rise to a serious limitation, namely a 2D SG model cannot dynamically distinguish between cold and warm fronts. We therefore also need to use a 3D SG model to further investigate the fundamental frontal dynamics. In formulating a simple 3D SG model, we introduce a reference state that has uniform stratification with a zonal flow characterized by $\overline{u}(y,z), \overline{\phi}(y,z), \overline{\theta}(y,z), \theta_o(z)$ on a *f*-plane. The reference state is to satisfy the following relations: $0=-\overline{\phi}_y - f\overline{u}$, $0=-\overline{\phi}_z + \frac{g}{\theta_{oo}}\overline{\theta}$, $(\theta_o)_z = \frac{\theta_{oo}}{g}N^2$. The total field of each dependent variable of a 3D flow in this semi-geostrophic model has constituent parts defined as follows:

$$u_{total} = \overline{u}(y,z) + \hat{u}(x,y,z,t) + \tilde{u}(x,y,z,t)$$
$$v_{total} = \hat{v}(x,y,z,t) + \tilde{v}(x,y,z,t)$$
$$w_{total} = \tilde{w}(x,y,z,t)$$
$$\theta_{total} = \theta_o(z) + \overline{\theta}(y,z) + \hat{\theta}(x,y,z,t)$$
$$\phi_{total} = \phi_o(z) + \overline{\phi}(y,z) + \hat{\phi}(x,y,z,t)$$

where $\hat{u}$, $\hat{v}$ and $\overline{u}$ are the components of the flow in geostrophic balance with $\hat{\phi}$ and $\overline{\phi}$ respectively; $\overline{\theta}$ and $\hat{\theta}$ the potential temperature components in hydrostatic balance with $\overline{\phi}$ and $\hat{\phi}$ respectively; and $(\tilde{u}, \tilde{v}, \tilde{w})$ the ageostrophic velocity component. These notations are used in order to have the cleanest possible look for the governing equations. The set of governing equations for this SG system are

$$0 = -\hat{\phi}_x + f\hat{v}, \tag{2.16a}$$

$$0 = -\hat{\phi}_y - f\hat{u}, \tag{2.16b}$$

$$0 = -\hat{\phi}_z - \frac{g}{\theta_{oo}}\hat{\theta}, \tag{2.16c}$$

$$\tilde{u}_x + \tilde{v}_y + \tilde{w}_z = 0, \tag{2.16d}$$

$$\frac{D(\overline{u}+\hat{u})}{Dt} = f\tilde{v}, \tag{2.16e}$$

$$\frac{D(\hat{v})}{Dt} = -f\tilde{u}, \tag{2.16f}$$

$$\frac{D(\overline{\theta}+\hat{\theta})}{Dt}+\tilde{w}\frac{d\theta_o}{dz}=0, \tag{2.16g}$$

where

$$\frac{D}{Dt}=\frac{\partial}{\partial t}+u_{total}\frac{\partial}{\partial x}+v_{total}\frac{\partial}{\partial y}+w_{total}\frac{\partial}{\partial z}.$$

The bottom and top boundaries $z = 0, D$, as well as the lateral boundaries $y = 0$, $y_{\max}$ are rigid surfaces across which there is no mass flux. Thus, the boundary conditions are: (1) $\tilde{w}=0$ at $z = 0, D$, (2) $\hat{v}=0$, $\tilde{v}=0$ at $y = 0$, $y_{\max}$ and (3) all dependent variables are cyclical at $x = 0$, $x_{\max}$. As in the case of a 2D model, (2.16) cannot be applied in this form. It would be necessary to reformulate them to a more amenable form.

2.4.1 Geostrophic Coordinates

As noted in Chapter 1, until recently all analyses of frontal dynamics using a 3D SG model reported in the literature were performed with the use of the so-called geostrophic coordinates. Those coordinates were first suggested by Yudin (1955): $X=x+f^{-1}\hat{v}$, $Y=y-f^{-1}\hat{u}$, $Z=z$, $T=t$ where $\hat{u}$ and $\hat{v}$ are the unknown geostrophic velocity components. They have the advantage of transforming the set of governing equations that incorporate the geostrophic momentum approximation similar in form to those of a QG model. With additional approximation they would become identical (Hoskins, 1975). The underlying reason is that the coordinate transformation continually stretches a locality of cyclonic vorticity and compresses a locality of anticyclonic vorticity while nonlinear dynamics induces a progressive contraction of scale along a frontal zone. The self-stretching coordinates would disguise the presence of the higher harmonics, and thereby would only implicitly represent the dynamical role of the ageostrophic circulation. However, the advantage gained in using the geostrophic coordinates does not come without a price since the dependent variables would have to be transformed to highly convoluted forms (eq. (10) in Hoskins and Draghici, 1977). In contrast, it would be desirable to analyze frontal development in physical coordinates (x, y, z) without imposing additional restrictions.

2.5 Analysis of a 3D SG Model in Physical Coordinates

A recent paper shows how an analysis of a 3D SG model can be made in physical coordinates (Mak, 2014). The governing equations (2.16) are a set of time-dependent nonlinear partial differential equations that can only be solved numerically. As in the case of dealing with a 2D SG model, there are two approaches for formulating a prognostic equation with which one could get new values of the geostrophic circulation over one time-step forward by numerical integration. They are elaborated in the following two subsections.

2.5.1 Three-Dimensional SG Potential Vorticity

In light of (2.16a,b,c) we introduce a geostrophic streamfunction, $\hat{\psi} = \frac{\hat{\phi}}{f}$, such that $\hat{u} = -\hat{\psi}_y$, $\hat{v} = \hat{\psi}_x$ and $\hat{\theta} = \frac{f\theta_{oo}}{g}\hat{\psi}_z$, implying thermal wind relations $\hat{\theta}_x = \frac{f\theta_{oo}}{g}\hat{v}_z$ and $\hat{\theta}_y = -\frac{f\theta_{oo}}{g}\hat{u}_z$. Thus, $\hat{\psi}$ embodies the information about $\hat{u}$, $\hat{v}$ and $\hat{\theta}$. The invariant property of this SG model is the *semi-geostrophic potential vorticity* q (Hoskins, 1975):

$$\frac{Dq}{Dt} = 0, \tag{2.17}$$

where

$$q = (-v_z^* + f^{-1}J_{y,z}(u^*, v^*))\theta_x^* + (u_z^* + f^{-1}J_{z,x}(u^*, v^*))\theta_y^* \\ + (f + v_x^* - u_y^* + f^{-1}J_{x,y}(u^*, v^*))(\theta_{oz} + \theta_z^*). \tag{2.18a}$$

$J_{y,z}(A,B) = A_yB_z - A_zB_y$ stands for a Jacobian, where $(u^*, v^*) = (\overline{u} + \hat{u}\ , \hat{v})$ is the total geostrophic velocity. One approach for evaluating the new values of the geostrophic circulation is to first calculate the change of q for one time-step with (2.17). This is supplemented by solving (2.18a) for the corresponding geostrophic streamfunction ψ field associated with the newly obtained q field since (2.18a) is effectively an inhomogeneous partial differential equation for ψ. With appropriate boundary conditions, this would be a feasible but very challenging task because (2.18a) is a nonlinear partial differential equation.

It is pertinent to briefly consider what one would have to deal with at this point of an analysis if one were to perform an integration in geostrophic coordinates (X, Y, Z) as defined in Section 2.4.1. The definition of potential vorticity would take on a simpler but still nonlinear relation to the geostrophic potential function, Φ, namely:

$$\frac{f^2}{q}\Phi_{ZZ} + \Phi_{XX} + \Phi_{YY} - \frac{1}{f^2}J_{XY}(\Phi_X, \Phi_Y) = f^2. \tag{2.18b}$$

where $\Phi = \phi + \frac{1}{2}(\hat{u}^2 + \hat{v}^2)$ is the potential function in the new coordinates; i.e., $(\Phi_X, \Phi_Y, \Phi_Z) = (\phi_x, \phi_y, \phi_z)$. After advancing each time-step, one would have to solve (2.18b) for obtaining the corresponding Φ field at the new time-step. The past simulation studies using geostrophic coordinates typically consider a special case of uniform potential vorticity, q = constant for simplicity (Hoskins, 1976; Hoskins and Heckley, 1981; Schar and Wernli, 1993). Furthermore, the nonlinear term in (2.18b) is *neglected* in those studies for convenience when the corresponding Φ field is computed by inversion. This expedient step was taken so that one would only need to solve a linear equation instead of a nonlinear equation. It is justified by arguing that such a neglected nonlinear term is generally small. But it would no longer be small as the front becomes progressively intense. Indeed, the neglect of it was found to produce noticeable changes in the development of a model front (Snyder et al., 1991).

2.5.2 Three-Dimensional SG Tendency Equation

As in the case of integrating a 2D SG model, an alternative approach is to work with a *semi-geostrophic tendency equation* by combining (2.16e,f,g). This is derived in exactly the same manner as that applied in deriving the QG tendency equation. Taking a derivative of (2.16f) with respect to x and (2.16e) with respect to y, we subtract the latter from the former, which would give us an equation that governs the vertical component of vorticity. It is next combined with the z-derivative of (2.16g), yielding:

$$P_t = -(\bar{u}+\hat{u}+\tilde{u})P_x - (\hat{v}+\tilde{v})P_y - \tilde{w}P_z - \tilde{u}_x\hat{v}_x + \tilde{u}_y\hat{u}_x - \frac{f^2}{N^2}\tilde{u}_z\hat{v}_z$$

$$-\tilde{v}_x\hat{v}_y + \tilde{v}_y\hat{u}_y + \frac{f^2}{N^2}\tilde{v}_z(\bar{u}_z+\hat{u}_z) - \tilde{w}_x\hat{v}_z + \tilde{w}_y(\bar{u}_z+\hat{u}_z) - \frac{f^2}{N^2}\tilde{w}_z\psi_{zz}$$

$$+\tilde{v}_y\bar{u}_y + (\hat{v}+\tilde{v})\bar{u}_{yy} + \tilde{w}\bar{u}_{zy}, \tag{2.19}$$

$$P = \left(\frac{\partial^2}{\partial x^2} + \frac{\partial^2}{\partial y^2} + \frac{f^2}{N^2}\frac{\partial^2}{\partial z^2}\right)\psi. \tag{2.20}$$

P is identifiable with the 3D QG potential vorticity.

2.5.3 Method of Analysis of the 3D Ageostrophic Circulation

After a new ψ field is obtained, the corresponding new $(\tilde{u}, \tilde{v}, \tilde{w})$ field still needs to be determined before one can proceed to the next time-step. This constitutes a greater challenge in an application of an SG model, in sharp contrast to an application of a QG model. The ageostrophic velocity $(\tilde{u}, \tilde{v}, \tilde{w})$ is related to the geostrophic streamfunction ψ by two diagnostic equations apart from (2.16d). Those two equations are obtained by eliminating the time derivative terms between (2.16e) and (2.16g) and between (2.16f) and (2.16g) on the basis of the thermal wind relations. The results are

$$A_1\tilde{u} + B_1\tilde{v} + C_1\tilde{w} = F_1, \tag{2.21}$$

where

$$A_1 = \hat{v}_z\frac{\partial}{\partial x} - (f+\hat{v}_x)\frac{\partial}{\partial z}$$

$$B_1 = -(\bar{u}_z+\hat{u}_z)\frac{\partial}{\partial x} - \hat{v}_y\frac{\partial}{\partial z},$$

$$C_1 = \left(\frac{N^2}{f} + \psi_{zz}\right)\frac{\partial}{\partial x} - \hat{v}_z\frac{\partial}{\partial z}$$

$$F_1 = 2(\bar{u}_z+\hat{u}_z)\hat{v}_x + 2\hat{v}_y\hat{v}_z$$

$$A_2\tilde{u} + B_2\tilde{v} + C_2\tilde{w} = F_2, \tag{2.22}$$

where

$$A_2 = -\hat{v}_z \frac{\partial}{\partial y} - \hat{u}_x \frac{\partial}{\partial z}$$

$$B_2 = (\overline{u}_z + \hat{u}_z)\frac{\partial}{\partial y} + (f - \overline{u}_y - \hat{u}_y)\frac{\partial}{\partial z}$$

$$C_2 = -\left(\frac{N^2}{f} + \psi_{zz}\right)\frac{\partial}{\partial y} - (\overline{u}_z + \hat{u}_z)\frac{\partial}{\partial z}$$

$$F_2 = 2(\overline{u}_z + \hat{u}_z)\hat{u}_x + 2(\overline{u}_y + \hat{u}_y)\hat{v}_z$$

Equations (2.16d), (2.21) and (2.22) form a closed set of linear partial differential equations for three unknowns $(\tilde{u}, \tilde{v}, \tilde{w})$ whenever the geostrophic velocity field may be regarded as known. Attempting to directly solve (2.16d), (2.21) and (2.22) would require very large computer storage since we would need high resolution to represent the model fronts. We can get the solution with a more effective method.

2.5.4 Generalization of the Sawyer–Eliassen Equation and Formal Solution

Since the 3D ageostrophic velocity $\tilde{\mathbf{V}} = (\tilde{u}, \tilde{v}, \tilde{w})$ is a non-divergent vector, $\mathbf{V}$ is equivalent to the curl of another vector field $\mathbf{A} \equiv (A^{(1)}, A^{(2)}, A^{(3)})$ so that $\tilde{\mathbf{V}} = \nabla \times \mathbf{A}$. $\mathbf{A}$ is generally known as a **vector potential**. We would gain nothing in trying to determine $\mathbf{A}$ instead of $\tilde{\mathbf{V}}$ since they both have three components. However, we know that $\mathbf{A}$ is not unique because we may introduce a gauge transformation, $\mathbf{A} + \nabla\vartheta \rightarrow \mathbf{B}$ where ϑ is an arbitrary scalar function. We would have $\nabla\cdot(\nabla \times \mathbf{B}) = 0$ and can deduce $\tilde{V}$ from $\mathbf{B}$. A desirable ϑ is one that would make one of the components of $\mathbf{B}$ zero.

There are three desirable choices of ϑ. The first choice amounts to partitioning the zonal ageostrophic velocity into two parts, say $\tilde{u} = \tilde{u}^{(1)} + \tilde{u}^{(2)}$ such that $\tilde{u}_x^{(1)} + \tilde{v}_y = 0$ and $\tilde{u}_x^{(2)} + \tilde{w}_z = 0$. It follows that we may introduce a pair of ageostrophic potentials φ and η such that

$$\tilde{u} = -(\varphi_y + \eta_z), \ \tilde{v} = \varphi_x, \ \tilde{w} = \eta_x \tag{2.23a}$$

The vector potential associated with (2.23a) is $\mathbf{B} = (0, \eta, -\varphi)$. The second choice of ϑ amounts to partitioning the meridional ageostrophic velocity into two parts, say $\tilde{v} = \tilde{v}^{(1)} + \tilde{v}^{(2)}$ such that $\tilde{u}_x + \tilde{v}_y^{(1)} = 0$ and $\tilde{w}_z + \tilde{v}_y^{(2)} = 0$. We may then introduce an alternative pair of ageostrophic potentials φ and η such that

$$\tilde{u} = -\eta_y, \ \tilde{v} = (\eta_x - \varphi_z), \ \tilde{w} = \varphi_y, \tag{2.23b}$$

The vector potential associated with (2.23b) is $\boldsymbol{B} = (-\varphi, 0, -\eta)$. The third choice of ϑ amounts to partitioning the vertical velocity into two parts, say $\tilde{w} = w^{(1)} + \tilde{w}^{(2)}$

such that $\tilde{u}_x + \tilde{w}_z^{(1)} = 0$ and $\tilde{v}_y + \tilde{w}_z^{(2)} = 0$. It follows that we may introduce another two ageostrophic potentials η and χ such that

$$\tilde{u} = -\eta_z, \; \tilde{v} = -\chi_z, \; \tilde{w} = \eta_x + \chi_y. \tag{2.23c}$$

The vector potential associated with (2.23c) is $\mathbf{B} = (-\chi, \eta, 0)$. This choice of partitioning corresponds to the one mentioned by Hoskins and Draghici (1977) except that theirs is defined for geostrophic coordinates in the context of the correspondingly transformed ageostrophic velocity components. All three vector-potentials above describe the ageostrophic velocity field equally well. Each can be reduced to the classic SE equation for the special case of a straight front.

With the use of η and χ defined according to (2.23c), we can rewrite (2.21) as

$$\begin{aligned} & S_1\{\eta\} + R_1\{\chi\} = G\{\psi\} \\ & \text{where } S_1 = (f + \hat{v}_x)\frac{\partial^2}{\partial z^2} - 2\hat{v}_z\frac{\partial^2}{\partial x \partial z} + \left(\frac{N^2}{f} + \psi_{zz}\right)\frac{\partial^2}{\partial x^2} \\ & R_1 = \hat{v}_y\frac{\partial^2}{\partial z^2} + \left(\frac{N^2}{f} + \psi_{zz}\right)\frac{\partial^2}{\partial x \partial y} - \hat{v}_z\frac{\partial^2}{\partial y \partial z} + (\bar{u}_z + \hat{u}_z)\frac{\partial^2}{\partial x \partial z} \\ & G\{\psi\} = 2(\bar{u}_z + \hat{u}_z)\hat{v}_x + 2\hat{v}_y\hat{v}_z. \end{aligned} \tag{2.24}$$

Equation (2.22) can be similarly rewritten as

$$\begin{aligned} & S_2\{\chi\} + R_2\{\eta\} = H\{\psi\} \\ & \text{where } S_2 = (f - \bar{u}_y - \hat{u}_y)\frac{\partial^2}{\partial z^2} + 2(\bar{u}_z + \hat{u}_z)\frac{\partial^2}{\partial y \partial z} + \left(\frac{N^2}{f} + \psi_{zz}\right)\frac{\partial^2}{\partial y^2} \\ & R_2 = -\hat{u}_x\frac{\partial^2}{\partial z^2} + \left(\frac{N^2}{f} + \psi_{zz}\right)\frac{\partial^2}{\partial x \partial y} - \hat{v}_z\frac{\partial^2}{\partial y \partial z} + (\bar{u}_z + \hat{u}_z)\frac{\partial^2}{\partial x \partial z} \\ & H\{\psi\} = -2(\bar{u}_z + \hat{u}_z)\hat{u}_x - 2(\bar{u}_y + \hat{u}_y)\hat{v}_z. \end{aligned} \tag{2.25}$$

S_1, S_2, R_1, R_2 are second-order linear partial differential operators with variable coefficients involving derivatives with respect to all three independent variables. $G\{\psi\}$ and $H\{\psi\}$ are known functions of ψ. We explicitly write the dependence of the RHS upon ψ in (2.24) and (2.25) to emphasize the fact that $\hat{u}$ and $\hat{v}$ are geostrophic velocity components. At any time when ψ may be considered known, the RHS of (2.24) and (2.25) as well as the variable coefficients in those four operators would be known. In other words, the SE equation has been generalized to two coupled partial differential equations, namely (2.24) and (2.25), that govern two potentials needed for depicting the ageostrophic circulation in physical coordinates. $G\{\psi\}$ and $H\{\psi\}$ play the role of forcing functions in (2.24) and (2.25). It is noteworthy that if there is no basic zonal flow, $\bar{u} = 0$, and if the geostrophic velocity field is a monochromatic wave, say $\psi = cos(kx + ly + mz)$, $G\{\psi\}$ and $H\{\psi\}$would be zero because

$G\{\psi\} = 2(\bar{u}_z + \hat{u}_z)\hat{v}_x + 2\hat{v}_y\hat{v}_z = 0$ and $H\{\psi\} = -2(\bar{u}_z + \hat{u}_z)\hat{u}_x - 2(\bar{u}_y + \hat{u}_y)\hat{v}_z = 0$. Then there would be no ageostrophic circulation associated with such a flow. However, when there is a basic zonal flow and/or the geostrophic part of the flow consists of a wave and its higher harmonics, the forcing functions would have finite values. Then the ageostrophic circulation associated with them would be the solution of (2.24) and (2.25).

It should be emphasized that the geostrophic part of the velocity also influences the ageostrophic circulation through its presence in the coefficients of (2.24) and (2.25), S_1, S_2, R_1, R_2. That part of the coefficients would make it possible for those equations to become singular, and hence a frontal collapse. The boundary conditions for η and χ are:

(1) $\eta = \chi = 0$ at $z = 0, D$ so that we would have $\tilde{w}^{(1)} = \eta_x = 0$, $\tilde{w}^{(2)} = \chi_y = 0$ at those surfaces;
(2) η and χ are both cyclical at the zonal boundaries of the channel;
(3) $\chi = 0$ at $y = 0, Y$ so that $\tilde{v} = \chi_z = 0$ (we use the notation $y_{\max} \equiv Y$ from here on without risk of confusion); and
(4) $\eta_y = 0$ at $y = 0, Y$ a less restrictive constraint.

It is noteworthy that the differential operators R_1 and R_2 are actually identical since we have $\hat{u}_x + \hat{v}_y = 0$ for geostrophic velocity. However, we continue to make a distinction between them because the lateral boundary conditions for η and χ are different.

The inverse of any differential operator C is an integral operator C^{-1} characterized by a specific kernel. By definition, we have $C^{-1}\{C\{h\}\} = I\{h\}$ for any function $h(x, y, z)$ where I is the unit operator. We provisionally assume that the inverse of operator S_1 exists. This would be the case when it is an elliptic differential operator. Under such circumstance, we may obtain from (2.24):

$$\eta = S_1^{-1}G\{\psi\} - S_1^{-1}R_1\{\chi\}. \tag{2.26}$$

By substituting this expression of η into (2.12), we obtain an expression for χ solely in terms of ψ:

$$\chi = (S_2 - R_2S_1^{-1}R_1)^{-1}(H\{\psi\} - R_2S_1^{-1}G\{\psi\}). \tag{2.27}$$

Putting this expression back into (2.26), we would also get:

$$\eta = \left(S_1^{-1}G\{\psi\} - S_1^{-1}R_1(S_2 - R_2S_1^{-1}R_1)^{-1}(H\{\psi\} - R_2S_1^{-1}G\{\psi\})\right) \tag{2.28}$$

provided that the combined operator $(S_2 - R_2S_1^{-1}R_1)$ also has an inverse. At certain times during integration, such an assumption may not hold and would have a drastic consequence. It is noteworthy that these operators are not commutative and they must be performed in the particular sequential orders indicated above. Equations (2.27) and (2.28) constitute a *formal solution* of the generalized SE equations as two separate *functionals* of a supposedly known function ψ. Much work remains to be done before we would be in a position to learn about the structure of the ageostrophic velocity component in a quantitative manner. The evolution of ψ in turn partly depends on η and χ at each instant in accordance with (2.19). Therefore, the ψ, η and χ fields must be calculated in tandem one after another at each time-step. We will present an illustrative analysis of 3D frontal dynamics by applying the equations formulated above in Chapter 5.

2.6 Concluding Remarks

The geostrophic and ageostrophic velocity components are not on an equal footing in an SG model. The relationship between them is one between a "slave" and a "master." The change in the master variable (geostrophic velocity) is governed by a prognostic equation. The corresponding change in the slave variable (ageostrophic velocity) is to be diagnostically determined with the SE equation in a 2D SG setting or with its generalization in the form of two coupled equations in a 3D SG setting. The newly evolving geostrophic circulation in each case plays the role of "forcing" the ageostrophic circulation, which produces a positive feedback to the frontal development. More general than an *SG* model would be a *primitive equation* (PE) model, which invokes hydrostatic balance as the only dynamical constraint. An even more general model would be a non-hydrostatic model. The essence of such a general model and two applications of it in frontal studies will be discussed in Chapter 7.

In passing, it should be added that there exist SG equations in the context of a domain other than an *f*-plane. Such works rest upon a generalized definition of the coordinate transformation and the geostrophic balance to take into account a variable Coriolis parameter. They were introduced by Salmon (1985) with the use of Hamilton's principle. This idea was adopted in a formulation of the SG equations for a *beta-plane domain* (Magnusdottir and Schubert, 1990). It invokes the simultaneous use of isentropic and generalized geostrophic coordinates. That particular formulation of the SG theory has been further generalized to a *hemisphere domain* by introducing a spherical coordinate generalization of the geostrophic momentum approximation (Magnusdottir and Schubert, 1991). They verified that the resulting equations have the desirable properties of conservation of absolute angular momentum, total energy, potential vorticity, and potential pseudodensity.

The most general form of the SG equations applicable to a *global domain* was formulated from Hamilton's principle (Shutts, 1989). Two approximations are introduced directly into the Hamiltonian principle in such a way that time and particle label symmetries are preserved. First, the kinetic energy associated with the component of velocity parallel to the axis of rotation is neglected. Second, the Lagrangian rate of change of the wind and pressure gradient directions (when projected onto the equatorial plane) is small compared to twice the angular rotation rate of the system. It follows that the equations automatically have global conservation of energy and

Lagrangian conservation of potential vorticity. The resulting system of equations have been shown to embody all fundamental properties (eigenmodes) of the primitive equations with a high degree of accuracy for meteorologically relevant equivalent depths. Atmospheric fronts are associated only with synoptic-scale disturbances in mid-latitude. It would not be necessary to work with any of these more general versions of SG equations for the purpose of investigating fundamental frontal dynamics.

It is also possible to construct solutions of the incompressible SG equations for the special case of piecewise constant data with a geometric method. The idea is to map two points in the geostrophic space (X, Y, Z) to two adjacent points in the physical space (x, y, z). The mathematical details underlying this method are given in section 3.4 of the book by Cullen (2006). This method was first illustrated by Cullen and Purser (1984).

Recap

- Quasi-geostrophic (QG) dynamics is intrinsically inadequate for modeling fronts because the advective influence of the ageostrophic velocity component is missing. Geostrophic momentum approximation:

$$\frac{DV}{Dt} \approx \frac{DV_g}{Dt}; \frac{D}{Dt} = \frac{\partial}{\partial t} + V_{total} \cdot \nabla.$$

- V is the total horizontal velocity, V_g is the geostrophic velocity component.
- The geostrophic momentum approximation is incorporated in a semi-geostrophic (SG) model.
- The simplest possible model framework adequate for representing frontal dynamics is an SG model.
- 2D SG model:
  - A single prognostic equation for the geostrophic streamfunction ψ is the 2D SG tendency equation.
  - A diagnostic equation relating the ageostrophic potential χ to ψ is the Sawyer–Eliassen (SE) equation.
- The SE equation would no longer be an elliptic partial differential equation when potential vorticity has negative values, $q < 0$, somewhere in the domain.
- Potential vorticity q plays the same role in SG dynamics as stratification N^2 in QG dynamics.
- 3D SG model:
  - A prognostic equation for ψ is the 3D SG tendency equation.
  - Ageostrophic circulation is represented by two potentials : χ and η.
  - Generalized SE equations: two coupled diagnostic equations relating χ and η to ψ.
  - Formal analytic solutions are obtained for χ and η as two separate functionals of ψ.
- There exist generalized versions of the SG equations for a beta-plane, a hemisphere, or a global domain.

3

Construction of Two Initial States for Modeling Fronts

3.1 Introductory Remarks

An initial state in a simulation study is prescribed with a specific objective in mind. For all of the semi-geostrophic (SG) models considered in Chapter 2, we will introduce an initial state as

$$\textit{initial state} = (\textit{reference state}) + (\textit{initial departure state}).$$

The reference state is intended to have the physical elements that would drive the fronts subsequently. For it to be relevant in a study of the atmospheric fronts, the reference state should contain some of the essential characteristics of the mean winter atmospheric state. In the special case of a 2D model, we consider a reference state that has a zonal shear flow $\overline{u}(z)$, varying with height in thermal wind balance with a certain temperature distribution $\overline{\theta}(y,z)$. There is an additional temperature component associated with a stable stratification $\theta_o(z)$. The departure state consists of a geostrophic velocity $\hat{v}$ in the y-direction with a corresponding potential temperature $\hat{\theta}$ and an ageostrophic velocity field $(\tilde{u},\tilde{w})$. They vary only in x, z, and t. The notations are the same as those used in Chapter 2. Summing up, the total fields of the dependent variables at any instant in our simple notations are:

$$u_{total} = \overline{u} + \tilde{u}, \quad v_{total} = \hat{v}, w_{total} = \tilde{w}, \quad \theta_{total} = \theta_o + \overline{\theta} + \hat{\theta}.$$

It is desirable to have a weak initial departure state $(\tilde{u},\hat{v},\tilde{w},\hat{\theta})$ that would immediately intensify in time in the presence of the chosen reference state. Therefore, a suitable initial departure state for a frontal model simulation could be an unstable normal mode of the corresponding reference state. Since a perturbation in a model is governed by the linearized form of the model equations, the advection of momentum and heat by the velocity of the reference state is much stronger than that by the velocity in the initially weak departure state. Then the linearized governing equations of an SG model (with an additional weak assumption $\theta_{oz} \gg \overline{\theta}_z$) would be indistinguishable from those of a quasi-geostrophic (QG) model, namely:

$$\hat{v}_t + \overline{u}\hat{v}_x = -f\tilde{u},$$

$$\hat{\theta}_t + \overline{u}\hat{\theta}_x + \hat{v}\overline{\theta}_y + \tilde{w}(\theta_{oz}) = 0,$$

$$\tilde{u}_x + \tilde{w}_z = 0,$$

$$\hat{\theta}_x = \frac{f\theta_{oo}}{g}\hat{v}_z.$$

It follows that we will construct an initial state on the basis of a modal instability analysis of a corresponding linearized QG model. Nevertheless, such a disturbance in the corresponding nonlinear SG model may be expected to intensify immediately in time as well. It should be added that frontogenesis may also develop from a non-modal disturbance for a limited duration even when the reference state under consideration is modally stable.

In this chapter we construct two initial states associated with two distinctly different reference states used in a 2D model in Sections 3.2 and 3.3. They will be used in the two illustrative frontal analyses to be presented in Chapter 4 for the purpose of delineating two different fundamental aspects of 2D frontal dynamics. Those analyses may be thought of as representative examples.

3.2 Two-Dimensional Eady Model Instability Analysis

The basic state in a continuous model constructed by Eady (1949) is the simplest possible reference state suitable for investigating the essence of extratropical cyclogenesis per se. The model is a rotating adiabatic inviscid Boussinesq fluid on an f-plane bounded above and below by rigid horizontal surfaces. It is based on the *QG approximation*. The basic state consists of a zonal shear flow $\overline{u} = \lambda z$, a corresponding component in the potential temperature field $\overline{\theta} = -\frac{f\lambda\theta_{oo}}{g}y$ and a basic stratification $N^2 = \frac{g}{\theta_{oo}}\frac{d\theta_o}{dz}$ where f, g, λ, N, and θ_{oo} are external parameters. It would suffice to focus on 2D disturbances for the departure state, $\frac{\partial}{\partial y} = 0$, in a domain $-\infty < x < \infty, 0 \leq z \leq D$. Such a perturbation has a geostrophic velocity component $(0, \hat{v}, 0)$ with a related potential temperature $\hat{\theta}$ field in thermal wind balance and an ageostrophic velocity component $(\tilde{u}, 0, \tilde{w})$. We may introduce a disturbance geostrophic streamfunction $\psi(x, z, t)$ to jointly represent $\hat{v}$ and $\hat{\theta}$ such that $\hat{v} = \psi_x$ and $\hat{\theta}_x = \frac{f\theta_{oo}}{g}\hat{v}_z$. It is governed by a linearized QG potential vorticity equation, namely

$$\hat{q}_t + \overline{u}\hat{q}_x + \hat{v}\overline{q}_x = 0, \tag{3.1}$$

where $\hat{v} = \psi_x$. The QG perturbation and basic potential vorticity are $\hat{q} = \psi_{xx} + \frac{f^2}{N^2}\psi_{zz}$ and $\overline{q} = \overline{\psi}_{xx} + \frac{f^2}{N^2}\overline{\psi}_{zz}$ respectively. Since the basic potential vorticity of this model is zero, the perturbation PV is also zero everywhere according to (3.1). The boundary conditions are $\tilde{w}=0$ at $z = 0, D$, implying that the perturbation potential temperature at those surfaces is constrained according to the following thermodynamic equation

$$@\ z=0,D \quad \hat{\theta}_t + \bar{u}\hat{\theta}_x + \hat{v}\bar{\theta}_y = 0. \tag{3.2}$$

The ageostrophic velocity component is diagnostically related to ψ at all times. By the continuity equation, we may introduce an ageostrophic potential η so that $\tilde{u} = -\eta_z$ and $\tilde{w} = \eta_x$. It is governed by a so-called QG omega equation as

$$\eta_{xx} + \frac{f_o^2}{N^2}\eta_{zz} = \frac{2f_o\bar{u}_z}{N^2}\psi_{xx}, \tag{3.3}$$

subject to the boundary conditions

$$@\ z=0,D \quad \eta=0. \tag{3.4}$$

As all coefficients in the governing equations are independent of x and t, a normal mode solution would have the form of a wave $\psi=\varphi(z)\exp(i(kx-\sigma t))$ and $\eta=\chi(z)\exp(i(kx-\sigma t))$ where $i=\sqrt{-1}$, k, and σ are constants. The real part of these expressions may be regarded as the solution of the disturbance in physical space. The amplitude function φ must then satisfy

$$\left(-\frac{N^2k^2}{f_o^2}\varphi + \frac{d^2\varphi}{dz^2}\right) = 0, \tag{3.5}$$

with boundary conditions

$$\begin{aligned} &@\ z=0, \quad \sigma\varphi_z = -k\lambda\varphi \\ &@\ z=D, \quad \sigma\varphi_z = -k\lambda\varphi + k\lambda D\varphi_z. \end{aligned} \tag{3.6a,b}$$

Also the two amplitude functions are related by

$$\frac{d^2\chi}{dz^2} - \frac{N^2k^2}{f_o^2}\chi = -\frac{2\lambda k^2}{f_o}\varphi, \tag{3.7}$$

with boundary conditions

$$@\ z=0,D, \quad \chi=0. \tag{3.8}$$

The horizontal wavenumber k can have any value, but the corresponding σ and φ must be such that (3.6) would be satisfied.

We measure horizontal distance, vertical distance, time and temperature in units of $L=\frac{ND}{f_o}$, D, $\frac{N}{f_o\lambda}$ and Θ, respectively. L is the radius of deformation. The horizontal velocity, vertical velocity, geostrophic streamfunction and ageostrophic potential are in units of $\frac{N}{f\lambda L}$, $\frac{N}{f\lambda D}$, $\frac{N}{f\lambda L^2}$ and $\frac{N}{f\lambda LD}$, respectively. The nondimensional parameters are $\hat{\lambda}=\frac{\lambda}{N}$, $\hat{k}=kL$. The nondimensional form of the hydrostatic balance is $\Gamma\hat{\theta}=\hat{\psi}_z$

with $\Gamma = \frac{g\Theta}{\lambda N D \theta_{oo}}$. The equivalent nondimensional domain is $0 \le x \le \frac{2\pi}{k}$ and $0 \le \hat{z} \le 1$. Relevant values of the parameters for the extratropical region are:

$$N = 10^{-2}\text{s}^{-1},\ \frac{\Theta}{\theta_{oo}} = 10^{-2},\ \lambda = 3 \times 10^{-3}\text{s}^{-1},\ D = 10^{4}\text{m, implying } L = \frac{ND}{f_o} = 10^{6}\text{m}.$$

The nondimensional form of (3.5) and (3.6a,b) is

$$\begin{aligned} &\left(-k^2\varphi + \frac{d^2\varphi}{dz^2}\right) = 0 \\ &\text{b.c.} \qquad @\ z = 0, \quad \sigma\varphi_{\hat{z}} = -k\varphi \\ &\qquad\quad\ \ @\ z = 1, \quad \sigma\varphi_{\hat{z}} = -k\varphi + k\varphi_{\hat{z}}. \end{aligned} \tag{3.9a,b,c}$$

This is an eigenvalue–eigenfunction problem. The eigenvalue σ only depends on k. Notice that the external parameters do not appear in these equations at all. The dependence of the eigenfunction φ on the external parameters is implicit in the units of length and time.

The general solution of φ is

$$\varphi = A\exp(kz) + B\exp(-kz). \tag{3.10}$$

A and B are integration constants. Substituting (3.10) into (3.9b,c), we get

$$\begin{aligned} &(\sigma + 1)A + (-\sigma + 1)B = 0 \\ &(\sigma + 1 - k)e^{k}A + (-\sigma + 1 + k)e^{-k}B = 0. \end{aligned} \tag{3.11a,b}$$

The condition for the existence of non-trivial values of A and B is that the determinant of the coefficients is zero, leading to a quadratic equation

$$\sigma^2 - \sigma k + [k\coth(k) - 1] = 0. \tag{3.12}$$

Thus there can only be two normal modes for each set of external parameters. The eigenfunction φ associated with each of these eigenvalues σ can be computed with the use of:

$$A = \frac{\sigma - 1}{\sigma + 1}B, \tag{3.13}$$

where B can have an arbitrary value.

The two roots of (3.12) for a particular wavenumber k of a baroclinic wave in this model are

$$\sigma = \frac{k}{2}\left(1 \pm \sqrt{1 - \frac{4\coth(k)}{k} + \frac{4}{k^2}}\right). \tag{3.14}$$

A wave would exponentially intensify in time if it is associated with an eigenvalue that has a complex value, $\sigma = \sigma_r \pm i\sigma_i$. That would be the case if its wavenumber is such that

$$1 - \frac{4\coth(k)}{k} + \frac{4}{k^2} < 0. \tag{3.15}$$

This condition is satisfied for sufficiently long waves (small enough k). In other words, there is no long-wave cutoff, but there is a short-wave cutoff for the instability in the Eady model. This type of instability is referred to as baroclinic instability. The growth rate of a 2D unstable wave under consideration is given by

$$\sigma_i = \frac{k}{2}\sqrt{\frac{4\coth(k)}{k} - \frac{4}{k^2} - 1}. \tag{3.16}$$

It should be emphasized that the condition for instability (3.15) only depends on k. Suppose we denote the nondimensional wavenumber at which the growth rate is maximum by k_{max} which, would have the same value under all parametric conditions. Since the dimensional horizontal length scale of the most unstable wave is $\frac{L}{k_{max}} = \frac{ND}{f_o k_{max}}$, it does not depend upon the baroclinic shear λ at all. This is a unique property of the Eady model. According to (3.10), the corresponding dimensional vertical length scale is equal to D/k_{max}, which does not depend upon the basic shear either. This is a related aspect of the Eady model. It warrants reiterating that the basic shear has no influence upon the selection of both the horizontal and vertical scales of the most unstable disturbance in the Eady model. This is a consequence of the extreme simplicity of the model. Indeed, we will see in the next two sections that the length scales of the dominant disturbance do depend on the basic shear when either the beta-effect and/or a more general basic flow are incorporated in a model.

The value of the growth rate of an Eady model, however, does depend upon the basic shear since $\sigma_i = \left(\frac{\sigma_i N}{f_o \lambda}\right)_{\text{dim}}$. The real part of the eigenvalue of an unstable disturbance depends on the parameter k as $\sigma_r = \frac{k}{2}$, meaning that the phase speed of *all* unstable Eady waves is equal to the basic flow at the mid-level, meaning the vertical mean of the basic flow. Forecasters refer to the level of a background flow equal to the propagation speed of a cyclone as the steering level of a cyclone.

The nondimensional governing equation for the amplitude function of the ageostrophic circulation is

$$\frac{d^2\chi}{dz^2} - k^2\chi = k^2 F, \tag{3.17}$$

with $F(z) = -2\lambda\varphi(z)$ subject to boundary conditions (3.8). The solution of (3.17) can be written in an integral form by applying the method of Green's function as

$$\chi = \int_0^1 G(z, z')F(z')dz'. \tag{3.18}$$

The Green's function $G(z, z')$ is defined by

$$\frac{d^2G}{dz^2} - k^2G = k^2\delta(z - z') \quad \text{for} \quad 0 \leq z \leq 1,$$
$$\text{b.c.} \quad G(0, z') = G(1, z') = 0. \tag{3.19}$$

where $\delta(z)$ is the Dirac delta function. It is found that

$$G(z, z') \equiv G_1(z, z') = \frac{k \sinh(k(z' - 1))\sinh(kz)}{\sinh(k)} \quad \text{for} \quad z' > z$$
$$G(z, z') \equiv G_2(z, z') = \frac{k \sinh(kz') \sinh(k(z - 1))}{\sinh(k)} \quad \text{for} \quad z' < z. \tag{3.20}$$

Thus, the solution for the amplitude function of the ageostrophic streamfunction is

$$\chi = -2\lambda\left(\int_0^z G_2(z, z')\phi(z')dz' + \int_z^1 G_1(z, z')\phi)(z')dz'\right)$$
$$= \frac{-\lambda k}{\sinh(k)}[P(z)\sinh(k(z - 1)) + R(z)\sinh(kz)]B, \tag{3.21}$$

where

$$P(z) = (1 - \hat{A})z + \frac{\hat{A}}{2k}(\exp(2kz) - 1) + \frac{1}{2k}(\exp(-2kz) - 1),$$

$$R(z) = (\exp(-k) - \hat{A} \exp(k))(1 - z) + \frac{\hat{A}e^k}{2k}(1 - \exp(2k(z - 1)))$$
$$+ \frac{e^{-k}}{2k}(1 - \exp(-2k(z - 1))),$$

$$\hat{A} = \frac{A}{B} = \frac{\sigma - 1}{\sigma + 1}.$$

Recall that B is an arbitrary constant associated with the geostrophic streamfunction. The results above enable us to compute the complete structure of a 2D baroclinic wave disturbance for a prescribed wavenumber k at, say, $t = 0$.

3.2.1 Instability Properties of Eady Model

Relevant values of the parameters for an extratropical region are: $N = 10^{-2}$ s^{-1}, $f_o = 10^{-4}$ s^{-1}, $\lambda = 3 \times 10^{-3}$ s^{-1}, $D = 10^4$ m, and $\frac{\Theta}{\theta_{oo}} = 10^{-2}$. Distance is measured in units of $L = \frac{ND}{f_o} = 10^6$ m and frequency in units of $\frac{\lambda f_o}{N} = 3 \times 10^{-5}$ s. In the interest of clarity, we now reuse the tilde notation to indicate nondimensional quantities. It follows that we would have $\tilde{\lambda} = 0.3$, $\Gamma = 0.33$. The growth rate of the Eady waves $\tilde{\sigma}_i$ as a function of wavenumber $\tilde{k}$ is shown in Figure 3.1. The most unstable wave has

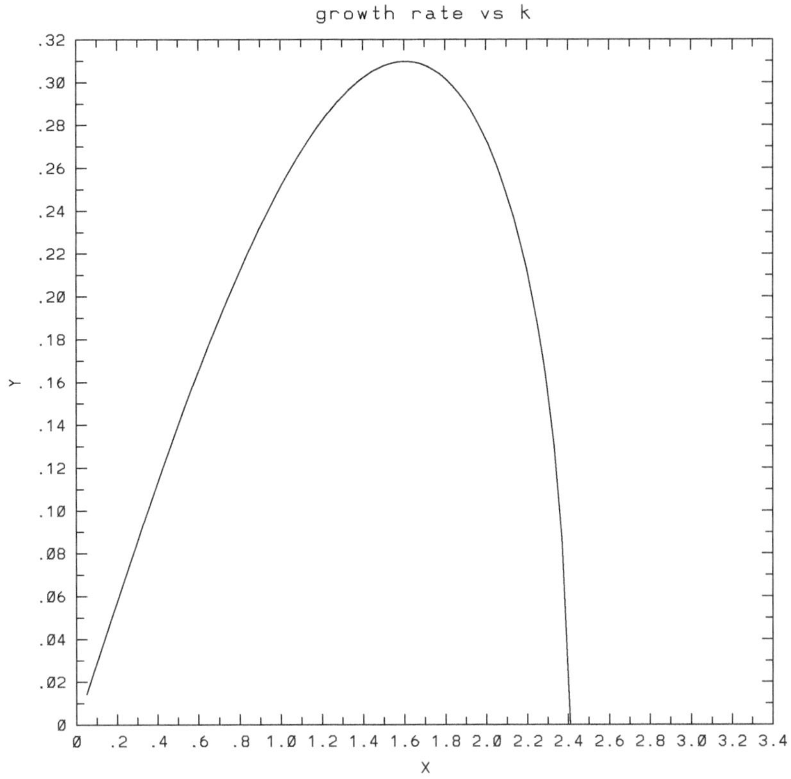

Figure 3.1. Variation of the growth rate of Eady waves $\tilde{\sigma}_i$ in units of $f_o\lambda/N = 3\times10^{-5}\ \mathrm{s}^{-1}$ with the nondimensional wavenumber $\tilde{k}$.

$\tilde{k}\approx 1.6$, so that its horizontal length scale is of the order of the *Rossby radius of deformation* $L=\frac{ND}{f_o}$. There is indeed no long-wave cutoff, but there is a short-wave cutoff in the Eady model at $\tilde{k}=\tilde{k}_c=2.4$. The maximum growth rate is about 0.31, corresponding to about $10^{-5}\ \mathrm{s}^{-1}$ or an e-folding time of ~ 1 day. It is noteworthy that the most unstable wavelength of the Eady model is independent of the basic shear. The phase speed of all unstable waves in the Eady model is eastward equal to the basic flow at the mid-level, $\frac{\sigma_r}{k}=\frac{\lambda D}{2}>0$.

All structural properties of the most unstable Eady wave are computed using $\tilde{k}=1.6$. $\tilde{B}=1$ would be a unit amplitude of the wave at the initial time. Recall that the geostrophic velocity component is measured in units of $\frac{f_o\lambda L}{N}=30\ \mathrm{m\ s^{-1}}$ respectively. Therefore, if we wish to prescribe a pertinent weak initial Eady wave, an appropriate value of $\tilde{B}$ would be $\tilde{B}=0.12$. Then the magnitudes of v and θ would be about 5 m s^{-1} and 2 K respectively. The structure of v and θ fields of the most unstable Eady wave are shown in Figure 3.2b,c. These two panels indicate that the geostrophic wind $\tilde{v}$ of this unstable wave has a characteristic westward vertical tilt and $\tilde{\theta}$ has a characteristic eastward tilt. The structure of the ageostrophic velocity components $\tilde{u}$ of this most unstable Eady wave is shown in Figure 3.2a.

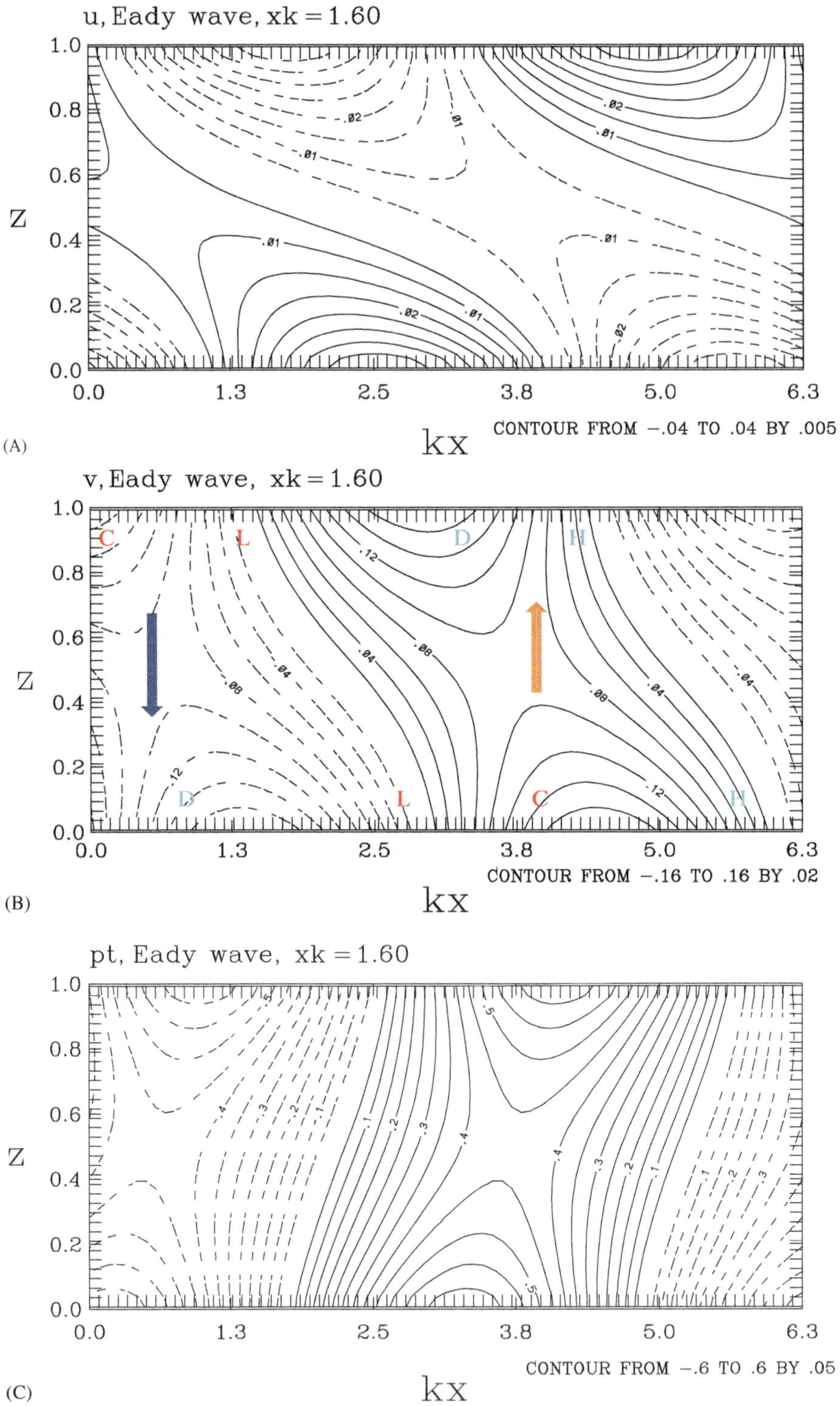

Figure 3.2. Complete structure of the most unstable Eady wave. (a) The ageostrophic velocity component in x-direction, $\tilde{u}$. (b) The meridional velocity component $\hat{v}$ in units of $f\lambda L/N = 30\ \mathrm{m\ s^{-1}}$. (c) The potential temperature field $\hat{\theta}$ in units of $\Theta = 3$ K. Additional labels added to panel (b): L (low pressure), H (high pressure), C (horizontal convergence), D (horizontal divergence), Arrows (max. and min. vertical velocity).

The amplitudes of $\tilde{u}$ and $\tilde{w}$ are about 1.2 m s^{-1} and 0.5 cm s^{-1} respectively. The horizontal convergence is $-\tilde{u}_x$. The geostrophic wind is zero and hence the streamfunction is minimum at the surface and location $kx \sim 2.82$. It is noteworthy that there is maximum horizontal convergence at the surface slightly to the east of maximum surface temperature. This low-level convergence supports an ascending motion. The maximum surface convergence is found at $kx \sim 4.12$, and there is divergence slightly to the west aloft over this location.

The relative structure of all properties is helpful for us to interpret the physical nature of the instability mechanism. Any instability mechanism entails a positive feedback process. We have added labels to Figure 3.2b to help visualize the relative structures of all properties. The instability mechanism can be interpreted from several complementary points of view. First, let us see how and why the surface pressure of this disturbance would deepen during intensification. The crucial feature of this amplifying Eady wave is its westward vertical tilt. Precisely because of such a tilt, there is an upper-level divergence stronger than the low-level convergence over a region of surface low pressure. Likewise, there is an upper-level convergence greater than the low-level divergence over a region of surface high pressure. Consequently, while the minimum pressure becomes lower, the maximum pressure would simultaneously become higher as the wave intensifies. At the same time, in the presence of planetary vorticity a low-level convergence tends to increase surface relative vorticity. In other words, this hydrostatically and dynamically balanced disturbance has the necessary overall structure for the continual intensification.

A consideration of the energetics of the disturbance would give us a complementary perspective of the intensification mechanism. We note that the warm (cold) sector is highly correlated with the southerly (northerly) flow. Hence, this unstable wave transports heat northward, enabling it to convert some of the basic available potential energy to wave available potential energy. Also the ascent in this vertical motion is located on the average in the warm sector, with descent in the cold sector. Rising warm air and simultaneously sinking cold air signifies a conversion of some of its available potential energy to kinetic energy. Consequently, both the velocity field and the temperature field would be amplified together in time. These are two traditional interpretations of baroclinic instability. We may add in passing that a counterpart *decaying Eady wave* would be characterized by an *eastward vertical tilt* for this basic flow in the northern hemisphere. The two interpretations discussed above are also applicable to a stable mode. In light of such vertical tilt, the decay of the wave stems from a counterpart, but negative, feedback process.

3.2.2 Essence of Baroclinic Instability in the Eady Model

The essence of the instability can be most succinctly explained from the perspective of potential vorticity dynamics in spite of the fact that the disturbance potential vorticity in this case is identically zero. The line of reasoning is a little bit more

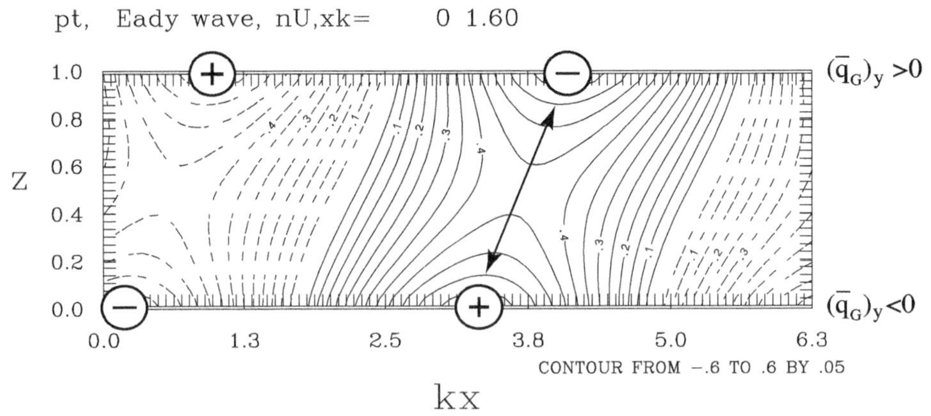

Figure 3.3. Indication of the equivalent generalized potential vorticity anomalies on the bounding surfaces (circles with + or −) in a plot of the potential temperature field of the most unstable Eady mode together with the characteristic of the basic gPV on the bounding surfaces.

abstract. Recall that the remaining prognostic equations of the model are the thermodynamic equation at the bottom and top bounding surfaces subject to the constraint $\tilde{w} = 0$, namely:

$$\hat{\theta}_t + \hat{v}\bar{\theta}_y = 0 \qquad \text{at } z = 0$$
$$\hat{\theta}_t + \bar{u}\hat{\theta}_x + \hat{v}\bar{\theta}_y = 0 \quad \text{at } z = D.$$

We recast these equations with the concept of generalized potential vorticity q_G (referred to as gPV for short), elaborated by Mak (2011: section 3.9.2). The appropriate form is $q_G = \frac{fg}{N^2\theta_{oo}}\theta\,\delta(z)$ at $z = 0$ and $q_G = \frac{-fg}{N^2\theta_{oo}}\theta\,\delta(z - D)$ at $z = D$ for a Boussinesq fluid, where $\delta(z)$ is the Dirac delta function, defined by $\int_{0-}^{0+}\delta(z)dz = 1$. Hence, a warm (cold) temperature disturbance at $z = 0$ ($z = D$) may be thought of as a positive gPV anomaly. Also, we have $0 > \left(\frac{d\bar{q}_G}{dy}\right)_{z=0}$, $0 < \left(\frac{d\bar{q}_G}{dy}\right)_{z=D}$ and $-\left(\frac{d\bar{q}_G}{dy}\right)_{z=0} = \left(\frac{d\bar{q}_G}{dy}\right)_{z=D}$ for the case of a westerly basic shear. The change of sign of $\left(\frac{d\bar{q}_G}{dy}\right)$ across the domain tells us that the necessary condition for instability is satisfied. The elements of a gPV of the same sign in one-half of a sinusoidal wave on each bounding surface may also be thought of as *an equivalent point gPV anomaly* at the center of that segment analogous of the center of mass in a body of finite size. We indicate the equivalent gPV anomalies and the characteristic of the basic gPV at the bounding surfaces in a plot of the potential temperature perturbation of an unstable mode in Figure 3.3. The signature feature of an unstable mode is that an equivalent gPV anomaly at $z = D$ is located to the *east* of an equivalent gPV anomaly of opposite sign at $z = 0$ as highlighted by a thick line with arrows in Figure 3.3.

We may visualize a local intrinsic cyclonic circulation associated with each element of a positive gPV anomaly since we can always determine the former by piecewise inversion of the latter. Generalized potential vorticity anomalies therefore influence one another via advection of basic gPV. The strength of the influence of a specific gPV anomaly is a function of distance from it, being weaker as it becomes further away. Now let us consider the influence of a particular gPV anomaly on another gPV on the opposite bounding surface. As the horizontal wavelength of an unstable baroclinic wave is two orders of magnitude longer than the depth of the domain, an equivalent gPV anomaly is mostly influenced by the nearest equivalent gPV anomaly across the domain. In light of the configuration of the gPV pattern, the equivalent positive gPV anomaly at $z = 0$ gives rise to a negative advection of basic gPV at the location of the equivalent negative gPV anomaly at $z = D$. The latter would therefore be strengthened by the interaction. By the same token, we visualize an intrinsic anticyclonic circulation associated with a negative gPV. Likewise, the equivalent negative gPV anomaly at $z = D$ gives rise to a positive advection of basic gPV at the location of equivalent positive gPV anomaly at $z = 0$. In other words, these two gPV anomalies would mutually strengthen one another by taking advantage of the basic state.

The mutual impact of the two gPVs would be lasting only if they are at rest relative to one another. It is thus necessary also to consider the movement of the gPV elements in this disturbance. A gPV anomaly also influences the neighboring gPV anomalies on the same bounding surface. The advection of basic gPV by the circulation associated with a gPV anomaly at $z = 0$ tends to increase the gPV of its neighbor to its east and decrease the gPV to its west because we have $0 > \left(\frac{d\bar{q}_G}{dy}\right)_{z=0}$. This process then has the net effect of moving the whole wave pattern at $z = 0$ *eastward*. We denote that speed by c. $\left(\frac{d\bar{q}_G}{dy}\right)_{z=0}$ is therefore akin to a negative beta-effect on a Rossby wave. Hence, the edge wave at a bounding surface is *dynamically the same* as a propagating Rossby wave. By the same token, the advection of basic gPV by the PV anomalies at $z = D$ similarly tends to move the wave pattern there *westward* at a speed equal to $(-c)$. Additionally, this edge wave at $z = D$ is advected by the basic flow there, namely $U(D)$. In order for the whole wave pattern of a normal mode to move as a single entity, we must then have $U(D) - c = c$ yielding $c = \frac{U(D)}{2}$ in agreement with the analytic result of the instability analysis. All elements of gPV in this disturbance are said to be phase-locked. The mutual reinforcement of the gPV anomalies across the domain elaborated earlier would therefore be able to persist, leading to exponential intensification. This is the most succinct way of interpreting the dynamical nature of modal baroclinic instability in the Eady model. It may be referred to as *wave resonance mechanism*, which is perhaps more popularly known as a form of *IPV (isentropic potential vorticity) thinking*. The applicability of this interpretation of shear instability is, of course, not restricted to the Eady model, but is valid for shear instability in general.

3.3 Two-Dimensional Equivalent-Charney Model (e-Charney) Instability Analysis

In this section, our objective is to construct a disturbance most pronounced only at the bottom surface of a model. It is to be used as an initial disturbance in a frontal illustrative analysis in Section 4.3. Such an unstable baroclinic wave is akin to the so-called Charney modes of baroclinic instability. The Charney (1947) model is more general than the Eady model in several aspects. In particular, the Charney model incorporates the variation of the Coriolis parameter with latitude, the so-called beta-effect. Its domain consists of a troposphere below a virtually unbounded stratosphere. Its basic state has a baroclinic flow of linear shear in the tropospheric region with a barotropic flow in the stratospheric region aloft. The beta-effect alone greatly complicates the mathematical analysis because the beta parameter gives rise to nonzero values in the potential vorticity of the basic state at all levels. The past theoretical analyses of the Charney model yielded some qualitative information about its instability properties. Without the benefit of a modern computer, Charney (1947) could only deduce limited information about the condition of marginal stability and make inference of the approximate structure of a marginally unstable wave. Burger (1962) analytically deduced that the basic flow in the Charney model has exponentially unstable modes for all wavenumbers, except at a set of isolated values of wavenumber. Kuo (1973) reported numerical instability properties for an asymptotic condition of the external parameters. They are compatible with the theoretical findings. A meticulous discussion of these subtle mathematical properties of the instability of the Charney model is given by Pedlosky (1987: section 7.8). Green (1960, figure 3) numerically found two distinct branches of unstable modes in the Charney model. Nowadays one branch is referred to as the Charney modes and the other branch the Green modes. Green pointed out the necessity of taking judicious steps in undertaking numerical computations with the Charney model as-is.

Our intent is to design a *Charney-like model* that could be straightforwardly used for getting robust instability properties of the baroclinic modes under the influence of the beta-effect. Such a model has a top boundary at a level $z = D - \Delta$ which is comparable to the tropopause height. The basic flow has a constant shear at all levels in this model domain. In light of the notion of wave resonance mechanism for shear instability elaborated in Section 3.2.2 (aka IPV thinking), we deliberately render the top bounding surface dynamically passive by simply incorporating next to it an infinitesimally thin layer ($\lim \Delta \to 0$) in which the shear asymptotically decreases to zero, $\lim_{z \to D} \left(\frac{d\bar{u}}{dz} \right) = 0$. For consistency, the basic meridional gradient of the temperature must also asymptotically decrease to zero in that layer. *In other words, the stratospheric region of the originally Charney model is reduced to an infinitesimally thin layer.* The e-Charney model is schematically shown in Figure 3.4. It should be noted that this treatment is also compatible with one particular result of Charney's

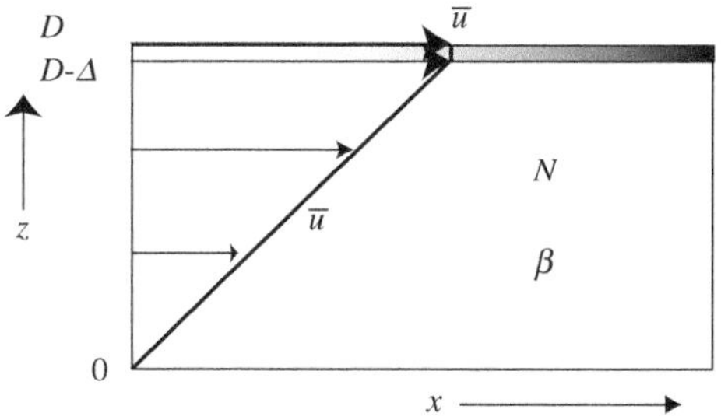

Figure 3.4. Schematic of the basic state in the e-Charney model.

investigation, specifically elaborated by him, that "the magnitude or direction of the zonal wind at very high levels in the stratosphere, say above 20 km, is of little consequence in the determination of the [unstable] wave velocity."

If there were no beta parameter in this model, the suppression of basic shear in the thin layer next to the top boundary would reduce the model to a slightly modified Eady model. But such a flow would be dynamically stable. The presence of beta-effect together with the basic baroclinicity, however, would restore the possibility of having instability because there are now nonzero values in the basic potential vorticity, $\overline{q}=\beta y$. It means that zonally propagating PV anomalies in the interior of the domain are possible, arising from their advection of planetary vorticity. These are in essence Rossby waves. A particular PV anomaly of a specific wavenumber at an interior level with a favorable structure can positively interact with a counterpart gPV anomaly at the bottom boundary (edge wave) in a sustained manner. Such mutual reinforcement could then give rise to baroclinic instability. We refer to this model as an equivalent-Charney model (*e-Charney model* for short). The subtle and crucial role played by the beta parameter in baroclinic instability was not known in Charney's time. He chose to incorporate it in his model simply because it seemed to be a necessary parameter for the upper-air waves, although it had been established to be so only for the wave motions in a barotropic atmosphere by Rossby. Since we purposely render the top boundary surface dynamically passive in the wave instability mechanism, the amplitude of unstable waves can be expected to be largest at the bottom surface.

In addition to the nondimensional quantities introduced in Section 3.1, we need to define one more nondimensional beta parameter, $\tilde{\beta}=\frac{\beta N^2 D}{f_o^2 \lambda}$. We nondimensionalize all other quantities as we did with the Eady model. The *nondimensional form* of the perturbation potential vorticity equation and the related boundary conditions are then:

$$
\begin{aligned}
&\left(\frac{\partial}{\partial t}+z\frac{\partial}{\partial x}\right)(\psi_{zz}+\psi_{xx})+\beta\psi_x=0 \\
&@\, z=0, \quad \psi_{zt}=\psi_x \\
&@\, z=1, \quad \psi_{zt}=-\psi_{zx}.
\end{aligned}
\tag{3.22}
$$

The corresponding nondimensional ageostrophic potential is governed by the QG omega equation:

$$\begin{aligned}&\left(\eta_{zz}+\eta_{xx}\right)=2\lambda\psi_{xx}\\&\text{at } z=0,1 \quad \xi=0.\end{aligned} \tag{3.23}$$

Although the beta parameter does not appear explicitly in (3.23), $\tilde{\eta}$ nevertheless feels its impact through the structure of $\tilde{\psi}$.

A complete normal mode solution consists of $\psi=\varphi(z)\exp(i(kx-\sigma t))$ and $\eta=\chi(z)\exp(i(kx-\sigma t))$. The two amplitude functions are then governed by

$$\begin{aligned}&(-\sigma+kz)(\varphi_{zz}-k^2\psi)+k\beta\varphi=0\\&@\, z=0, \quad -\sigma\varphi_z=k\varphi\\&@\, z=1, \quad -\sigma\varphi_z=-k\varphi_z\end{aligned} \tag{3.24}$$

and

$$\begin{aligned}&(\chi_{zz}-k^2\chi)=-2\lambda k^2\varphi\\&\text{at } z=0,1 \quad \chi=0.\end{aligned} \tag{3.25}$$

Equation (3.24) shows that the solution of φ depends on the external nondimensional parameter β as well as k. The nondimensional shear also appears explicitly in (3.25), which diagnostically relates χ to φ. The general solution of φ may be written as

$$\varphi=AF(z)+BG(z), \tag{3.26}$$

where F and G are hypergeometric functions. A and B are integration constants. In order to satisfy the boundary conditions, we require

$$\begin{aligned}&AF'(1)+BG'(1)=0\\&\sigma=\frac{-k(AF(0)+BG(0))}{AF'(0)+BG'(0)}=\frac{k(F(0)G'(1)-G(0)F'(1))}{-F'(0)G'(1)+G'(0)F'(1)}.\end{aligned} \tag{3.27a,b}$$

It is virtually impossible to analytically determine the roots of σ for a given set of parameters that would satisfy (3.27b) in general because the hypergeometric functions, F and G, themselves have complicated parametric dependence on σ. It is, however, not difficult at all to get the eigenvalues and eigenfunctions by solving the differential equation (3.24) with a numerical scheme. The eigenvector $\varphi(z)$ associated with a particular eigenvalue can in turn be used to solve (3.25) for the corresponding solution of $\chi(z)$.

It is instructive to first consider the special case of $\beta = 0$. The RHS of (3.27b) would become particularly simple because the potential vorticity of a normal mode, $\psi_{xx}+\psi_{zz}$, in such a case is necessarily zero everywhere. Then there would be no dependence on σ in F and G for they would become exponential functions, $F=e^{kz}$

and $G=e^{-kz}$. The eigenvalue is then $\sigma=\coth k$, which has a real value for any value of k, confirming that there can be no baroclinic instability without the beta-effect in this e-Charney model. This result highlights the pivotal role played by the beta-effect in the instability mechanism. Furthermore, the relation between the integration constants is $A=e^{-2k}B$. It follows that the solution of the geostrophic streamfunction of a neutral normal mode is $\psi=B(e^{-2k}e^{kz}+e^{-kz})\cos(kx-\sigma t)$, which monotonically decreases away from the surface boundary without any vertical tilt. This is also necessarily so from the energetic and dynamic points of view.

As far as the author can tell, the parameter of crucial importance to the problem under consideration, namely $\tilde{\beta}=\frac{\beta N^2 D}{f_o^2 \lambda}$ is unnamed in the literature. *It would be fitting to call it the Charney number, in honor of J.G. Charney's pioneering contribution in solving the baroclinic instability problem with the influence of beta-effect.* The Charney number may be thought of as the ratio of two *vertical length scales*, $\tilde{\beta}=\frac{D}{h_*}$ where $h_*=\left(\frac{\lambda f_o^2}{N^2}\right)/\beta$, which is the vertical length scale over which advection of planetary vorticity balances the stretching of planetary vorticity by the vertical velocity field. The vertical length scale of an unstable disturbance in this model is not necessarily equal to the vertical extent of the domain D. If the density scale height is large enough, the vertical length scale of the most unstable wave could be h_* instead implying that it is in general dependent on the vertical shear in contrast to the unstable modes in the Eady model. We may equally well think of the Charney number as the ratio of two *horizontal length scales*, $\tilde{\beta}=\frac{L}{L_*}$ where $L_*=\left(\frac{\lambda f_o}{N}\right)/\beta$. In other words, the horizontal length scale needs not be the Rossby radius of deformation, $L=\frac{ND}{f_o}$ and could be L_* instead, meaning that the horizontal length scale of the most unstable disturbance in general would also be dependent on the basic shear in contrast to the unstable modes of the Eady model. These considerations imply that there can be multiple branches of unstable modes in the e-Charney model as in the Charney model. This anticipation must be verified by numerical computation.

It is noteworthy that $\tilde{\beta}$ would be inversely proportional to the vertical shear λ for a given set of values in all other external parameters. Therefore, we may think of the $(\tilde{\beta},\tilde{k})$ plane as the inverse-shear versus wavenumber plane in the parameter space. Our specific task is to ascertain how the instability properties of our e-Charney model vary on the $(\tilde{\beta},\tilde{k})$ parameter plane, whereby to illustrate the extent to which the e-Charney model captures the physics of the Charney model.

3.3.1 Instability Properties of the e-Charney Model

For a set of relevant atmospheric values of the parameters $N=10^{-2}$ s^{-1}, $f_o=10^{-4}$ s^{-1}, $\lambda=3\times10^{-3}$ s^{-1}, $D=10^4$ m and $\beta=1.6\times10^{-11}$ m^{-1} s^{-1}, we would have $\tilde{\lambda}=0.3$ and $\tilde{\beta}\approx0.5$. For convenience, we again drop the tilde notation in the following discussion. It is found that there is at most one pair of complex-conjugate eigenvalues for a wide range of the parameters β and k. Since the instability solely depends on the Charney number β for a given wavenumber k, let us begin by examining such dependence for a synoptic-scale wave $k = 1.0$. The eigenvalues for $0 \leq \beta \leq 10.0$ are shown in Figure 3.5. Panel (b) reveals three non-overlapping branches of unstable modes in this range. The first branch exists between $\beta > 0$ and $\beta \approx 2.0$. The maximum growth rate is about 0.36 for $\beta \approx 1.1$. The corresponding e-folding time is about $1. \times 10^5$ s (~1 day). The result that the growth rate is largest for an intermediate value of β (Figure 3.5b) suggests that the beta-effect per se can also have a stabilizing consequence. The nature of this subtle dependence of the instability upon β will be further elaborated later. The second branch of unstable modes prevails between $\beta \approx 2.1$ and $\beta \approx 8.1$. The maximum growth rate of this branch is only about 0.08 at $\beta \approx 6.0$, which is four times smaller than that of the first branch for a basic shear about six times weaker. The third branch of unstable modes starts at $\beta \approx 9.0$, with even smaller growth rate values. This general decrease in growth rate in the three branches with increasing β is reasonable since β is inversely proportional to the basic shear. The important feature is that there are multiple maximum growth rates at certain values of β for a given value of k.

Figure 3.5a shows that the phase velocity of the unstable modes in the first branch decreases almost linearly with increasing β. The phase velocity of the second branch varies mildly with β between two zero values through a maximum value of ~0.17. The third branch of unstable modes has similar properties with even smaller values. Figure 3.5a,b for our e-Charney model are the counterpart of figure 12 in Kuo's (1973) paper for the eigenvalues of the Charney model and are compatible with Burger's (1962) qualitative deduction.

It would be instructive to get a broad view of how the eigenvalue varies with β and k by repeating the computation for meaningful combinations of these parameters. The variations of the phase velocity $\mathrm{Re}\{\sigma\}/k$ and growth rate $\mathrm{Im}\{\sigma\}$ of the modal disturbances are depicted in Figures 3.6a,b respectively for the ranges $0 < k \leq 10$ and $0 < \beta \leq 30$. In particular, Figure 3.6b reveals isolated maximum growth rate values on the (β, k) plane. When we focus on the variation of the instability properties with β for a fixed value of k in this plot, say $k = 1$, we identify three branches of unstable modes as previously seen in Figure 3.5. When we focus on the variations of the instability properties with k for a fixed value of β, say $\beta = 6.0$, we see two branches of unstable modes. One branch exists at $k < 3.5$ (long waves) and another branch at $k > 3.5$ (shorter waves), suggesting that the set of long unstable waves is the counterpart of the Green modes and the set of shorter unstable waves the Charney modes. The Charney model has an amazing property in that, except for an infinite set of discrete wavenumbers, all wavenumbers are unstable. This property was deduced analytically by Burger (1962). The numerical result of the e-Charney model shown in Figure 3.6b is fully compatible with the analytic

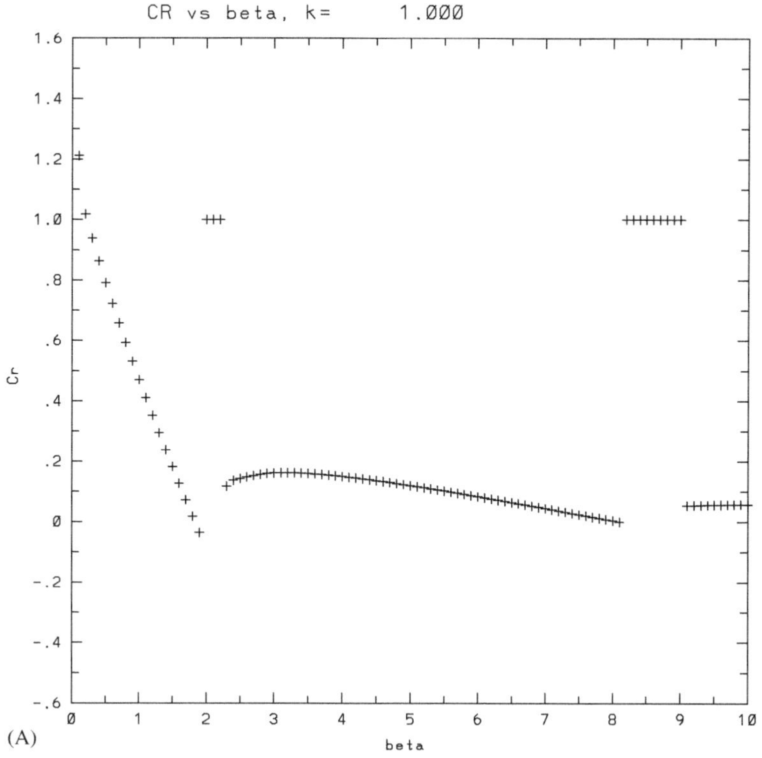

(A)

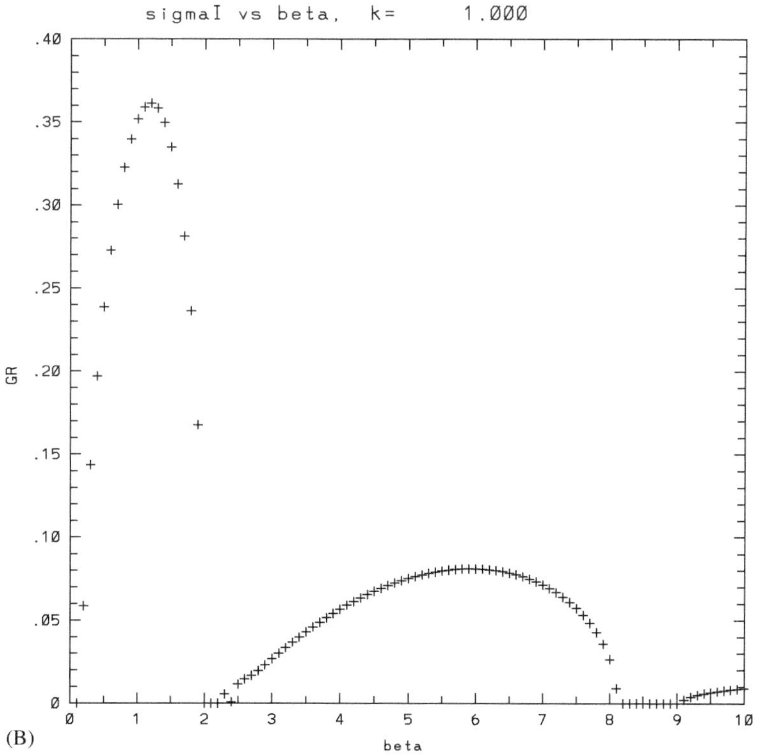

(B)

Figure 3.5. Variation of (a) the phase velocity, $\mathrm{Re}\{\sigma\}/k$ in units of $\lambda D = 30$ m s^{-1} and (b) the growth rate, $\mathrm{Im}\{\sigma\}$ in units of $\frac{f_o\lambda}{N} = 3 \times 10^{-5}$ s^{-1} of a normal mode with β for $k = 1.0$ in the e-Charney model.

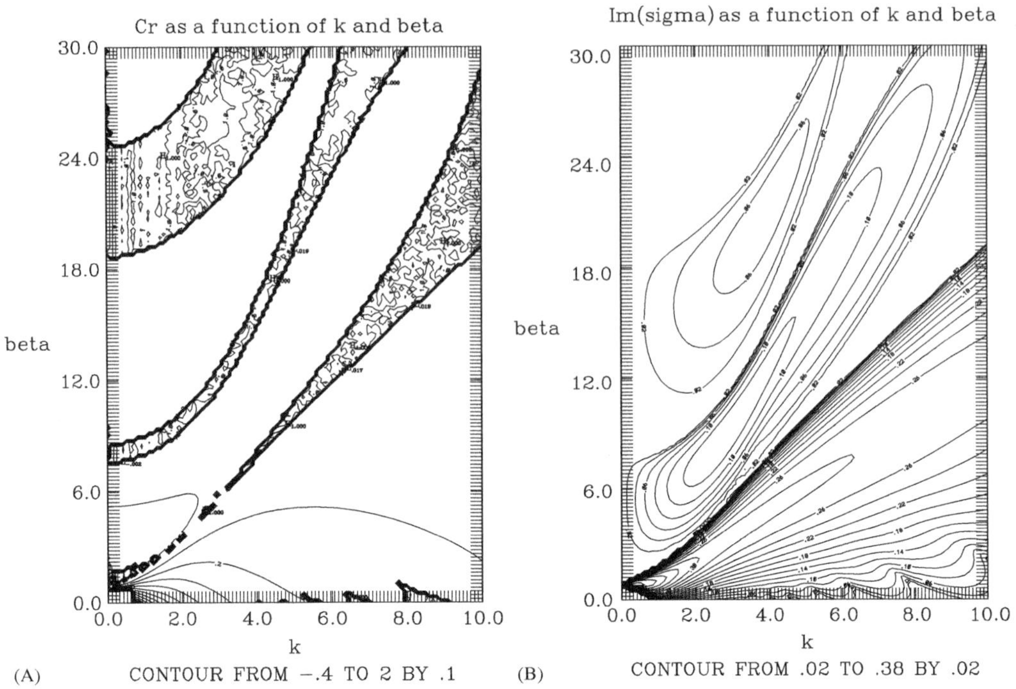

Figure 3.6. Variation of the instability properties of the normal modes in the e-Charney model with the zonal wavenumber k and $\beta = \frac{\beta_{\text{dim}} DN^2}{\lambda f_o^2}$. (a) Phase velocity, $\text{Re}\{\sigma\}/k$ in units of $|\lambda|D = 30\text{ m s}^{-1}$ and (b) growth rate, $\text{Im}\{\sigma\}$ in units of $\frac{f_o\lambda}{N} = 3\times10^{-5}\text{ s}^{-1}$.

property. The physical basis of this property will shortly become apparent when we discuss the essence of the instability in the Charney model and hence e-Charney model in Section 3.3.2.

The overall picture documents the existence of multiple non-overlapping branches of unstable modes separated by a set of curves on the (β, k) plane. The higher branches in this model exist for larger values of β. The modes of the higher branches have progressively smaller growth rates. For example, the maximum growth rate of the first three branches are about 0.4, 0.13 and 0.07 respectively. Within each branch, the growth rate is largest for an optimal value of β for a given wavenumber k. The phase velocity of the neutral modes is effectively 1.0, which is equal to the basic flow at the top of the domain.

The three branches of unstable modes may be expected to differ in their vertical structures. Figure 3.7 shows a comparison of the structure of the ψ field of one unstable normal mode from each of the three branches. The parameters of those modes are (β, k) = (0.5, 1.0), (7.0, 2.0) and (18.0, 3.0). It shows that the modes of a higher branch have shorter vertical scales. The mode of the lowest branch has maximum amplitude at the bottom surface $z = 0$ and extends to about the model mid-troposphere. The mode of the second branch has a node at $z \approx 0.4$. The mode of

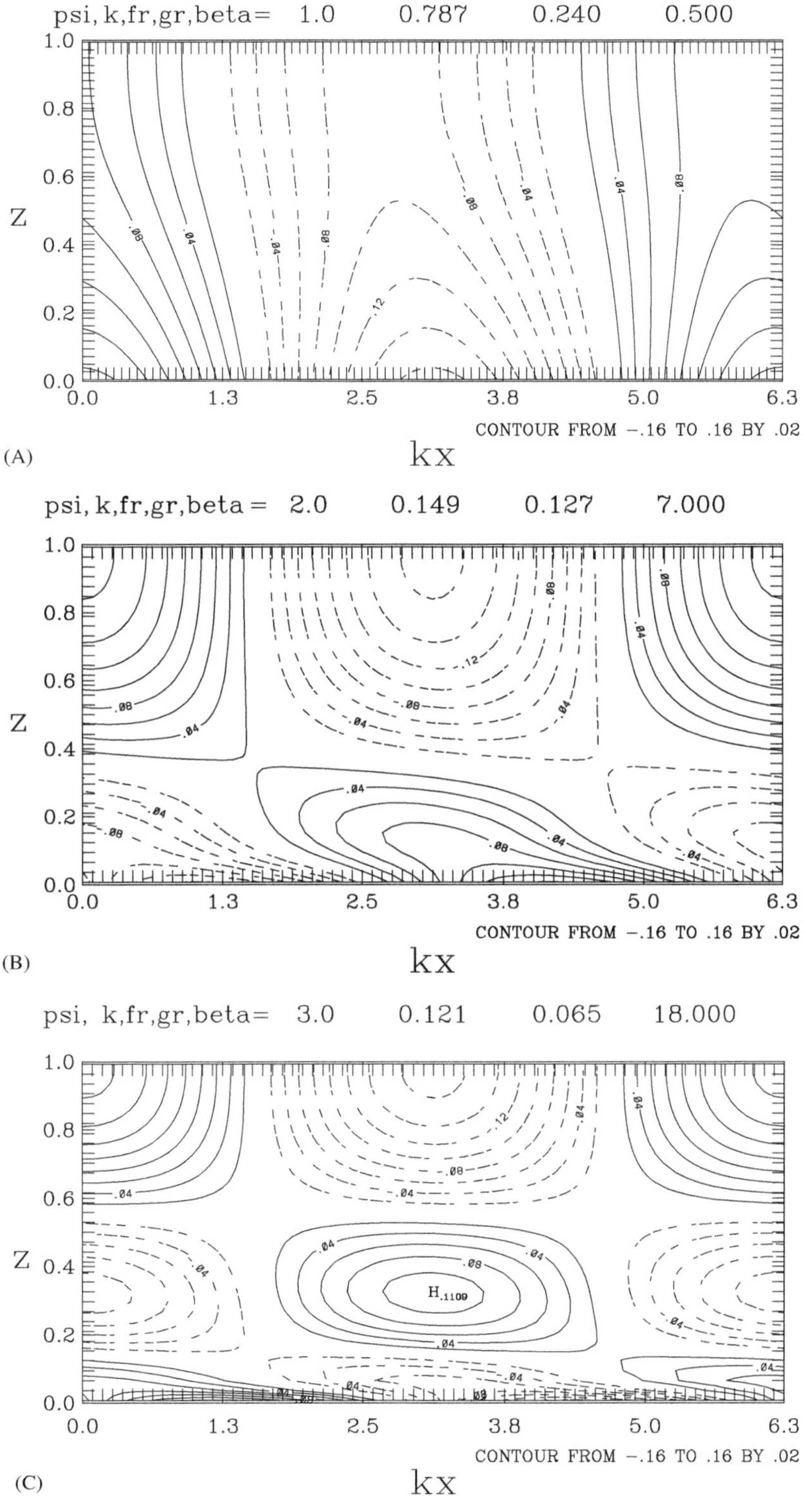

Figure 3.7. Structure of ψ of a representative unstable mode in the three branches of the e-Charney model, for (a) $(\beta, k) = (0.5, 1)$, (b) $(\beta, k) = (7, 2)$, and (c) $(\beta, k) = (18, 3)$ in units of $|\lambda| ND^2/f_o = 3 \times 10^7$ m^2 s^{-1}.

the third branch has two nodes, one at $z = 0.18$ and the other at $z = 0.6$. It is noteworthy that each of these unstable modes has a baroclinic structure next to the bottom surface and has a virtually barotropic structure above. This result is compatible with the structure of the unstable Charney mode for an asymptotic parametric condition (Kuo, 1973). The barotropic structure of these modes in the upper part of the domain stems from the vanishing baroclinicity in the infinitesimally thin layer attached to the top bounding surface of this e-Charney model. Larger values of the Charney number $\left(\tilde{\beta} = \frac{\beta N^2 D}{f_o^2 \lambda}\right)$ correspond to the cases of smaller values of vertical shear λ. It is natural, then, that an unstable mode of a higher branch associated with a weaker vertical shear would have a smaller growth rate and a shallower baroclinic structure next to the bottom surface. The lowest branch indeed corresponds to the Charney modes, whereas the higher branches correspond to the Green modes. In light of the results presented in these two subsections, we may conclude that only the unstable modes of the first branch (Charney modes) are relevant to cyclogenesis in the atmosphere.

We next closely examine the structural properties of a synoptic-scale unstable mode in the first branch of this e-Charney model. It is close to being the most unstable mode for a representative value of β. The parameters for this mode are $(\beta, k) = (1.0, 1.0)$. Its nondimensional growth rate is 0.35 and the phase velocity is 0.47. Figure 3.8a shows that the geostrophic velocity of this mode is mostly confined below $z = 0.4$, with its amplitude being strongest at the bottom boundary. It has a westward vertical tilt in the lower half of the troposphere. The normal mode has been normalized so that its maximum amplitude of the easterly surface is about 5 m s^{-1}. The structure of its potential temperature disturbance θ field is shown in Figure 3.8b. Its amplitude is also strongest at the surface, with a value of about 2.5 K. It has an eastward vertical tilt. The steering level of this mode is slightly below the level of mean basic zonal velocity. This unstable wave transports heat northward because of the positive correlation between its v and θ. It therefore can release some basic potential energy. We use the eigenfunction of φ to solve (3.25) for the corresponding ageostrophic streamfunction χ, which is shown in Figure 3.8c. There is positive correlation between the vertical motion and the potential temperature. This wave therefore simultaneously converts some wave potential energy to wave kinetic energy. The overall structure of this mode is similar to a Charney mode and has all the essential characteristics of an unstable baroclinic wave.

3.3.2 Essence of the Instability in e-Charney Model

We can interpret the essence of the instability in the e-Charney model from the perspective of potential vorticity dynamics. The potential vorticity field, $q = \psi_{xx} + \psi_{zz}$, is nonzero stemming from the presence of the beta-effect (Figure 3.8d). This result

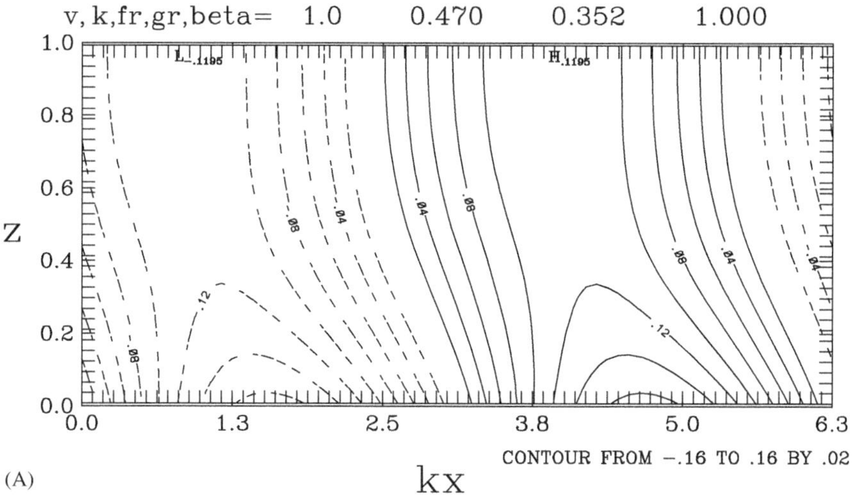

(A)

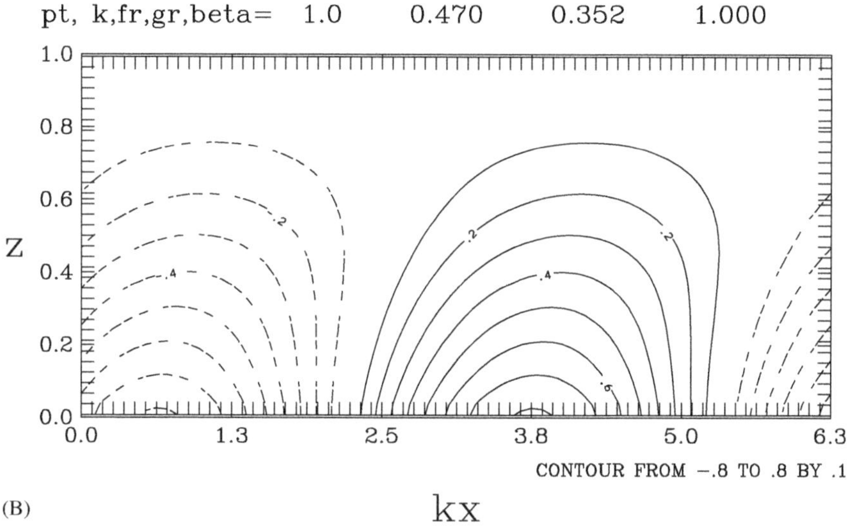

(B)

Figure 3.8. Structure of the unstable mode for $(\tilde{\beta}, \tilde{k}) = (1.0, 1.0)$ in the e-Charney model: (a) the geostrophic along-front velocity field in units of $f_o \lambda L/N = 30$ m s^{-1}; (b) the potential temperature field in units of Θ = 3 K; (c) the ageostrophic streamfunction in units of $f_o \lambda DL/N = 3 \times 10^5$ m^2 s^{-1}; and (d) the potential vorticity field in units of $f_o \lambda /N = 3 \times 10^{-5}$ s^{-1}, with indication of the equivalent PV anomalies (P and N), basic PV, gPV anomalies (circles with + or –) and basic gPV.

is computed using 21 levels to depict the vertical domain, $0 \le z \le 1$. The q field calculated with more levels, e.g., 35, has a slightly more noisy structure because the truncation error in ψ_{zz} of q is greater when the grid distance is reduced. The positive PV anomalies may be represented by an equivalent positive point PV anomaly. It is indicated by the letter P at $(xk,z) \approx (0.8, 0.5)$ in Figure 3.8d. Likewise, the equivalent negative point PV is indicated by the letter N at $(xk,z) \approx (3.9, 0.5)$.

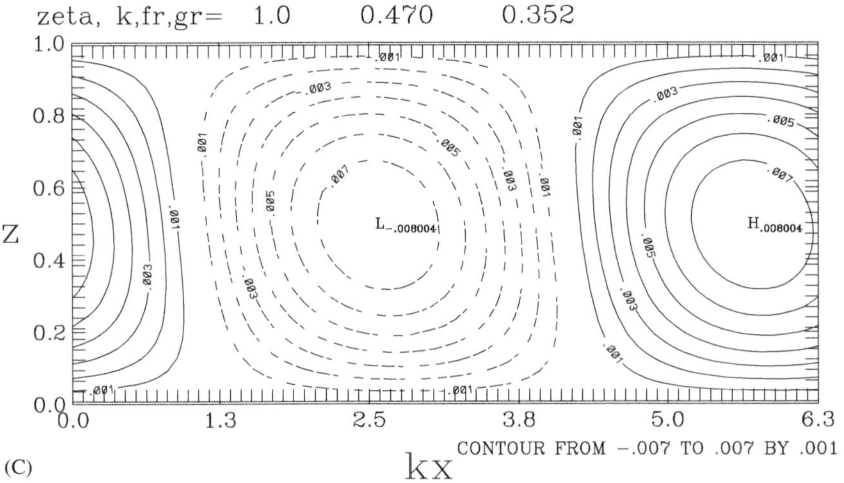

(C)

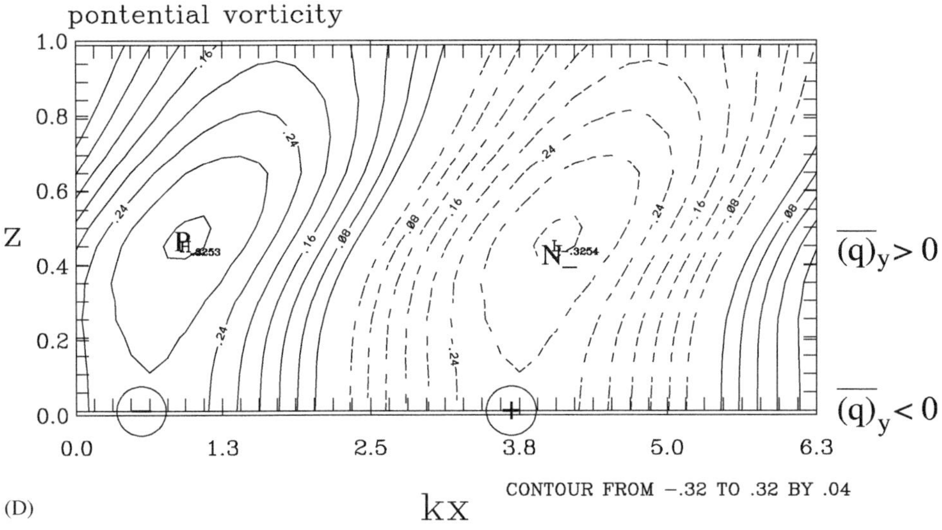

(D)

Figure 3.8. (*Cont.*).

We now invoke the notion of gPV, q_G, discussed in Section 3.1.3. For a Boussinesq fluid, we have $q_G = q + \frac{f_o g}{N^2 \theta_{oo}} \theta\, \delta(z)$ at the surface boundary $z = 0$. The equivalent gPV anomalies are indicated by circles with + or –. A signature feature of an unstable mode is that the equivalent negative PV anomaly, N, is located to the *east* of the equivalent gPV anomaly. We may visualize an intrinsic flow associated with each PV or gPV anomaly. The intrinsic flow at the surface associated with N would give rise to positive advection of basic gPV at the location of the circle with a +. This process therefore tends to strengthen the equivalent positive gPV anomaly at $z = 0$. Likewise, the equivalent positive gPV tends to strengthen the equivalent negative PV due to the advection of the basic PV associated with the beta-effect. The same is true for the interaction between the equivalent positive PV anomaly in the interior and the equivalent negative gPV at $z = 0$. In other words, there is mutual reinforcement of all PV anomalies. Since the elements of the gPV anomaly as a whole are phase-locked in a normal mode, the mutual reinforcement of the interior PV anomalies and the surface gPV elaborated above would continue, resulting in exponential growth of the normal mode. This is the most succinct interpretation of the dynamical nature of the baroclinic instability in this Charney model of finite depth. In light of this physical interpretation of the instability in the e-Charney model, the absence of short-wave cutoff in this model is readily understandable. It is always possible to have a PV anomaly associated with a wave of any wavelength existing at certain vertical levels because the beta-effect can resonantly interact with a surface PV anomaly associated with the surface baroclinicity. Such a wave can therefore be unstable.

The vertical length scale of an unstable mode in each branch is the vertical extent of its baroclinic structure next to the bottom surface. Recall that such a scale of the second branch is distinctly shorter than that of the first branch. That is what we should anticipate since according to our earlier interpretations of the physical meaning of the Charney number, both the vertical length scale and the horizontal length scale of unstable disturbances in the e-Charney model are dependent on the vertical shear. Not surprisingly, the interpretation of the essence of instability elaborated above for the first branch is applicable to the higher branches as well. After all, they are all baroclinic unstable modes. In each case, the instability stems from mutual reinforcement between a gPV anomaly at the bottom surface and a PV anomaly in the interior close to the bottom surface.

In conclusion, all features in the numerical results of the unstable modes of this e-Charney model have counterparts in the unstable modes of the Charney model. It is justifiable to suggest that the e-Charney model captures the essence of baroclinic instability of the Charney model.

Recap

- *Initial state = (reference state) + (initial departure state).*
- A pertinent initial state of a specific SG model for investigating frontal dynamics is an unstable normal mode of the reference state of a counterpart QG model.
- Two-dimensional Eady model:
  - All instability properties are known analytically.
  - No long-wave cutoff; there exists short-wave cutoff.
  - Horizontal and vertical length scales of the most unstable mode are independent of the basic shear.
  - Unstable modes of a westerly basic shear flow have a characteristic westward vertical tilt.
  - Essence of instability: mutual reinforcement of the generalized potential vorticity at the two bounding surfaces (wave resonance mechanism).
- Two-dimensional equivalent-Charney model:
  - Incorporates beta-effect, with a dynamically passive top surface.
  - Charney number $\tilde{\beta} = \frac{\beta N^2 D}{f_a^2 \lambda}$, basic shear λ; other notations are standard.
  - Uniform vertical shear in a tropospheric layer capped by a dynamically passive top.
  - Straightforward to numerically determine instability properties.
  - Unstable for all wavenumbers except for a set of isolated values.
  - Non-overlapping multiple branches of unstable modes:
    - o lowest branch, the Charney modes;
    - o higher branches, the Green modes; shorter vertical scale, longer horizontal scale.
  - Length scale of the most unstable mode of each branch depends on the basic shear.
  - Essence of instability: wave resonance mechanism.

4

Two-Dimensional SG Frontal Dynamics

4.1 Introductory Remarks

In this chapter we present two illustrative analyses of frontal dynamics in the 2D semi-geostrophic (SG) setting. The two corresponding different reference states have been constructed in Chapter 3 for different objectives. The modal instability analyses of these reference states have already been presented in Chapter 3. A suitably chosen unstable mode in each case is used as an initial disturbance for the corresponding SG frontal simulation. The algorithm elaborated in Section 2.3.3 is applied in these analyses. We will be in a position to gain some insight into the intrinsic frontal dynamics in each case. The chapter ends with a discussion of the limitations of a 2D model setting for investigating frontal dynamics.

4.2 Two-Dimensional SG Frontal Eady Model Analysis

The reference state adopted for this particular analysis is the familiar Eady flow in a rotating Boussinesq fluid on an f-plane. It consists of $\overline{u}=\lambda z$, $\overline{\theta}=a-\frac{f\theta_{oo}\lambda}{g}y$, with $(\theta_o)_z=\frac{\theta_{oo}}{g}N^2$. This is a *minimal frontal model* in the sense that if it were simplified any further, no disturbance with frontal characteristics would form. The most unstable baroclinic wave of this reference state is used as an initial disturbance. The unique aspect of this set-up is that both the bottom and top surfaces would play an indispensable role in the intensification of the baroclinic wave and hence frontal formation. The structure of the most unstable Eady wave in terms of its nondimensional variables $\tilde{v}, \tilde{u}, \tilde{w}$ and $\tilde{\theta}$ is shown again in Figure 4.1 for easy reference, although it has been presented and extensively discussed in Section 3.2. Apart from a vertical tilt, each field is symmetric about the mid-level.

The integration is performed using a time-step of 15 minutes. The domain of this model ($0\leq kx\leq 2\pi$, $0\leq z\leq 1$) is represented by 75 grid points in the zonal direction and 35 grid points in the vertical direction. At $t=0$ the western boundary of the warm sector is located at $kx=1.8$ and has an eastward vertical tilt (Figure 4.1b). This initial eastward tilt gradually tips toward the vertical and eventually changes to

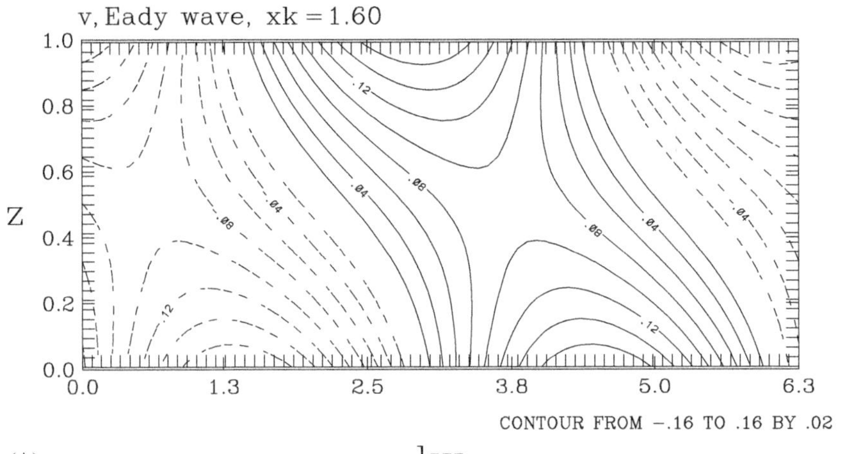

(A)

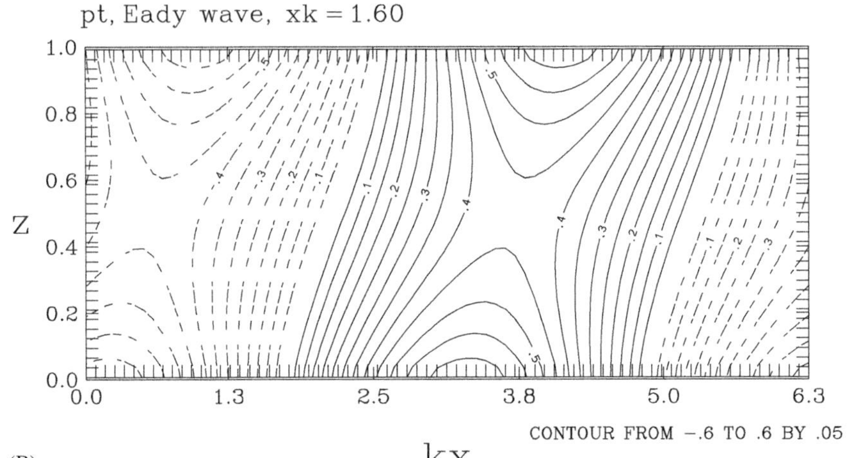

(B)

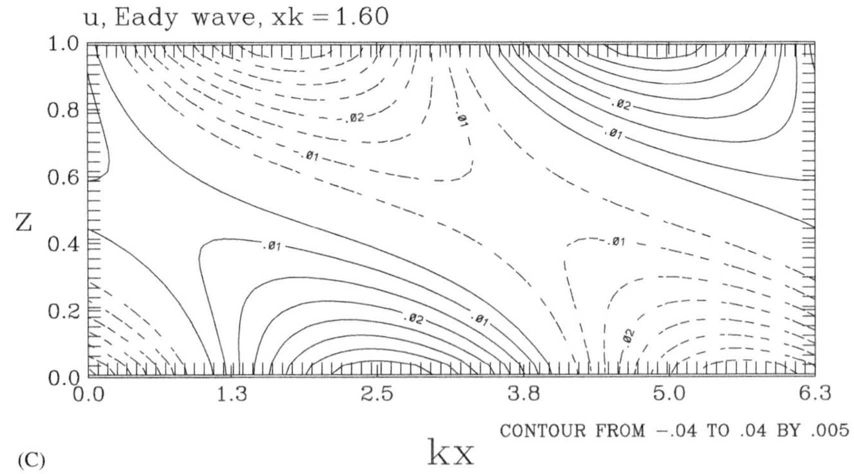

(C)

Figure 4.1. Complete structure of the most unstable Eady wave: (a) the meridional velocity component in units of $f\lambda L/N = 30\,\mathrm{m\,s^{-1}}$; (b) the potential temperature field in units of $\Theta = 3\,\mathrm{K}$; (c) the meridional velocity component in units of $f\lambda L/N = 30\,\mathrm{m\,s^{-1}}$; and (d) the vertical velocity field in units of $f\lambda D/N = 0.3\ \mathrm{m\,s^{-1}}$.

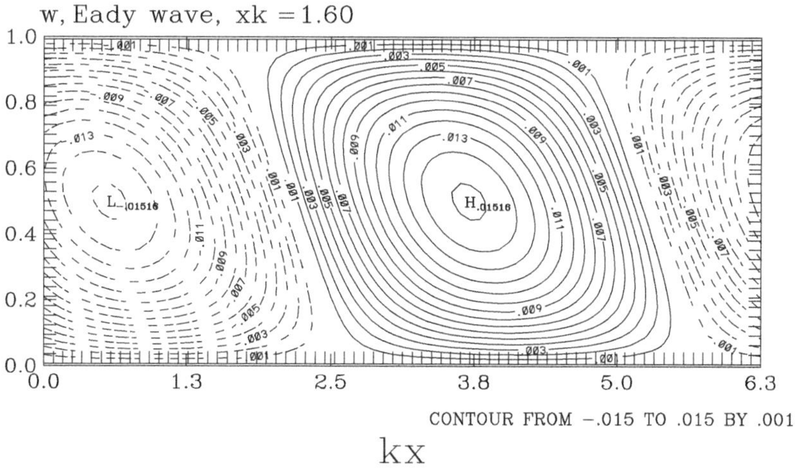

(D)

Figure 4.1. (*Cont.*).

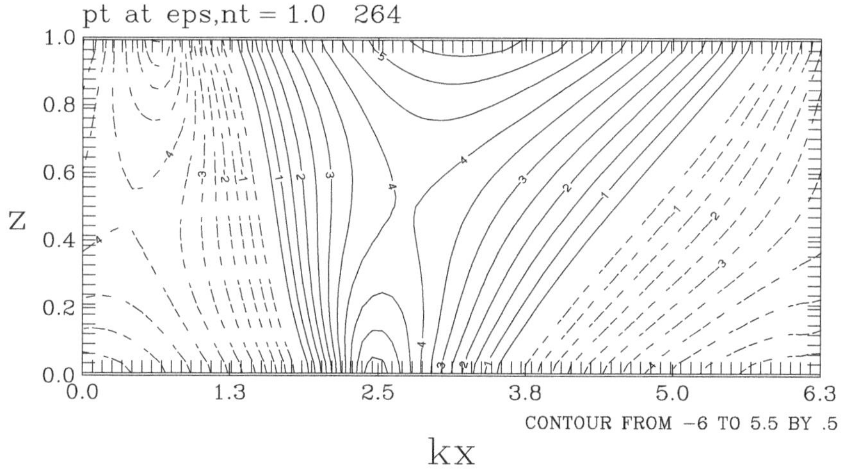

Figure 4.2. Distribution of the potential temperature disturbance in units of $\Theta = 3\,\mathrm{K}$ at the 264th time-step (day 2.75).

a westward tilt. A moderate front already appears at this boundary surface at $kx = 2\pi$ on day 2. In other words, the disturbance has propagated eastward by about 71percent of a wavelength in the first two days. The magnitude of the disturbance θ has increased from 0.6 to 3.3 in these two days (by almost five-fold). By day 2.75, this disturbance has moved further eastward by another 16 percent of a wavelength and intensified to 6.0. By this time, the area of warm anomalies has the shape of an upside down trapezoid. The area of cold anomalies changes to a counterpart zonally shifted trapezoid. The sector of warm anomaly has greatly contracted at the bottom surface and correspondingly stretched at the top surface. The reverse happens to the sector of cold anomaly, giving rise to a frontal boundary slanting westward between ($kx = 1.9$, $z = 0$) and ($kx = 1.5$, $z = 1$). This dramatic change in the structure of the disturbance potential temperature field θ after 2.75 days is shown in Figure 4.2.

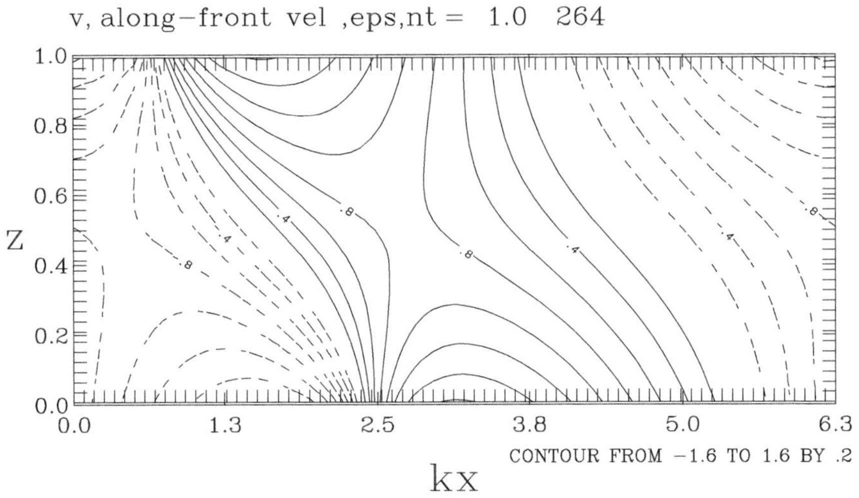

Figure 4.3. Structure of the geostrophic velocity at the 264th time-step (day 2.75) in units of $\lambda D = 30\,\mathrm{m\,s^{-1}}$.

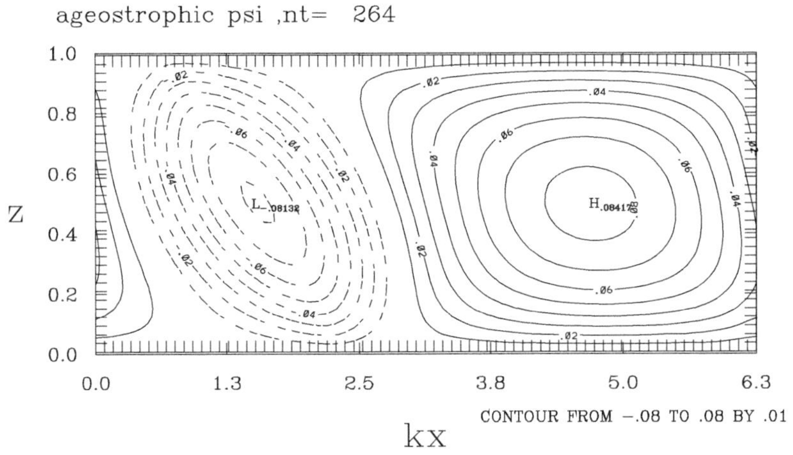

Figure 4.4. The ageostrophic streamfunction ξ field on day 2.75 in units of $\lambda D^2 = 3 \times 10^5\,\mathrm{m^2\,s^{-1}}$.

The frontal structures at the bottom and top bounding surfaces are equally pronounced in the corresponding geostrophic velocity field on day-2.75 (Figure 4.3). The westward tilt of v is greater than that of θ and its shear at the fronts is strong and cyclonic.

The scale contraction and stretching arises from the nonlinear dynamics. It is particularly evident in the ageostrophic potential field on day 2.75. It is shown in Figure 4.4. The two cellular ageostrophic circulations on the vertical plane give rise to a band of relatively narrow ascending motion between $kx \sim 2$ and $kx \sim 3.7$, centering at their common boundary. Its width is only about 22 percent of the domain. One cell is much narrower than the other. The maximum ascent is found to be about 4.5 cm s^{-1} at $(kx, z) = (2.5, 0.35)$. There is maximum descent of the same strength at $(kx, z) = (0.7, 0.65)$. The positive correlation between the vertical motion and warm anomaly indicates that potential energy is converted to kinetic energy, supporting the inexorable intensification of the whole disturbance throughout frontogenesis.

4.2.1 How and Why Do Fronts Develop in This Minimal Model?

The results above verify that 2D fronts do develop from an embryonic unstable baroclinic wave of a reference state. Recall that baroclinic instability in an Eady model may be understood as a process of mutual reinforcement between an edge wave at the bottom boundary and another edge wave at the top boundary (*wave resonance mechanism* for short). Those edge waves are associated with the meridional thermal gradient at the bounding surfaces of the reference state. From the perspective of generalized potential vorticity, these waves are dynamically Rossby waves. When the disturbance intensifies, the nonlinear advective effects of its ageostrophic flow component in this SG model become progressively significant. It generates the higher harmonics of the initial disturbance. These waves are dispersive and therefore their group velocity is a function of wavelength. In a 2D model setting, the group velocity of all zonal wavelengths is eastward and the shorter waves have slower group velocities. As the wave field intensifies, the longer wave components would progressively catch up with the shorter waves. The superposition of all spectral components of the wave ensemble consequently gives rise to the appearance of an intensifying frontal surface. Its frontal features are most pronounced at the bottom and top boundaries by virtue of the kinematic constraint.

The scale contraction of the ageostrophic cellular circulation to the west of the surface front and the corresponding scale-stretching of the cell of opposite direction to the east of the surface front, as seen in Figure 4.4, is an important part of the frontogenetic process. Deformation of the horizontal flow in this model is the same as the divergence of its ageostrophic flow component $((u_{total})_x - (v_{total})_y = \hat{u}_x)$. Hence, instead of invoking a horizontal deformation flow acting on its temperature field as a frontogenetic process, we may simply attribute that to the impacts of the horizontally convergent ageostrophic flow and the compensating vertical motion. The ageostrophic flow next to the bottom surface enhances the frontal thermal gradient toward the point of maximum convergence by bringing together fluid of contrasting temperature from the two sides. This convergence of fluid in turn supports an ascending motion along a slanting frontal surface. The ascending motion that brings the fluid from below extends the thermal contrast up to the top boundary and provides the mass flux in support of a divergent flow there. In this sense, the fronts at the two bounding surfaces owe their existence to different aspects of the overall process. We cannot state whether this surface front should be designated a cold front or a warm front. This is an intrinsic limitation of a 2D SG model.

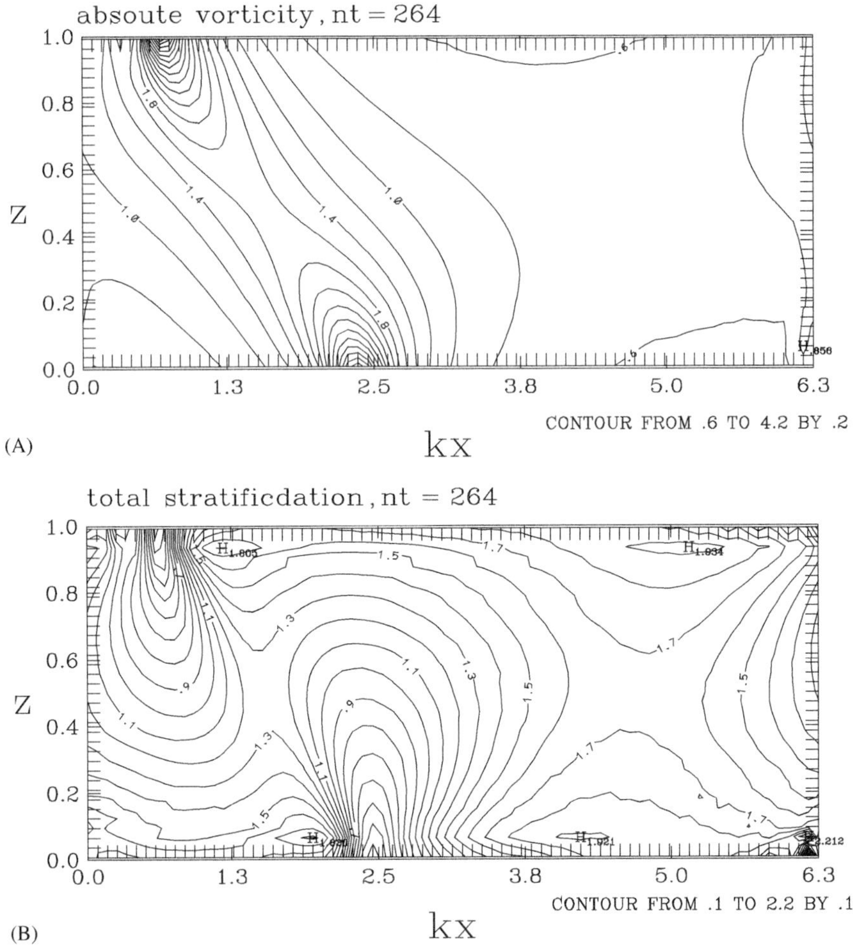

Figure 4.5. Distribution of the (a) absolute vorticity $(1+\lambda v_x)$ in units of $f_o = 10^{-4}\,\mathrm{s}^{-1}$; and (b) the total stratification $(1+\lambda\Gamma\theta_z)$ in units of $N^2 = 10^{-4}\,\mathrm{s}^{-2}$.

It is found that the Sawyer–Eliassen (SE) equation remains elliptic up to the 264th time-step (day 2.75), but the condition of ellipticity breaks down shortly afterward because the model resolution can no longer adequately resolve the fields by this point. Numerical noise overwhelms each field and the integration stops shortly afterward. This may be interpreted as the demise of the model fronts since this development has a clear physical basis. The absolute vorticity field and the total stratification field at that time are shown in Figure 4.5. Panel (a) shows a maximum value of the absolute vorticity of 4.2 at about $xk \approx 2.4$ on the surface boundary and at $xk \approx 0.7$ on the top boundary. Panel (b) shows a minimum value of stratification as small as 0.1 at this time. The thermal advection by the ageostrophic flow near the surface decreases (increases) the stratification to the east (west) of the front. By inference, the breakdown of the elliptic condition for the SE equation is attributable to symmetric instability. It limits how strong the front could be in an SG model. In the atmosphere, convection would occur as soon as the total static stability takes on a negative value. The convection would quickly adjust the stratification back to a neutrally stable configuration. When we apply dry convective adjustment to the model, the frontogenetic process indeed proceeds further and the fronts are found even sharper than what we saw in Figure 4.3.

4.2.2 Impacts of Dissipation

One would intuitively expect that dissipation, however one might represent it, would have the effect of limiting the minimum width and the maximum intensity of a front. Nakamura (1994) quantitatively examined the impacts of diffusion on frontogenesis in this 2D SG model by means of numerical simulations with both horizontal and vertical Laplacian types of diffusion. The presence of diffusion introduces additional viscous terms in the SE equation, thereby suppressing a breakdown of its elliptical condition. There would then be no frontal collapse in such a viscous model. Viscous effect becomes important shortly after 200 hours of integration. Figure 4.6 shows that at t = 200 h the evolution of the flow is similar to the inviscid solution, very comparable with Figures 4.2 and 4.3 obtained in our analysis. Pronounced fronts form at the bottom and top surfaces of the model.

Due to the influence of the diffusion, the model fronts do not continue to intensify toward a singularity, but instead begin to weaken shortly afterward. By t = 260 h, the two model fronts almost align vertically. The vertical tilt is subsequently reversed. Like we saw in Figure 4.4, there is great asymmetry between the two cells of ageostrophic circulation, as seen in Figure 4.7 at t = 240 h.

The circulation is found subsequently to undergo oscillations progressively toward a much weakened steady state, as documented in Figure 4.8. It particularly reveals that the model diffusion induces potential vorticity through the boundaries.

It should be added that Cullen (2006) reported multiple cycles of fronts in this same 2D *inviscid* model continuing to day 30 (see his figure 6.6), according to his calculation obtained with a geometric method. It is, however, unclear as to why the

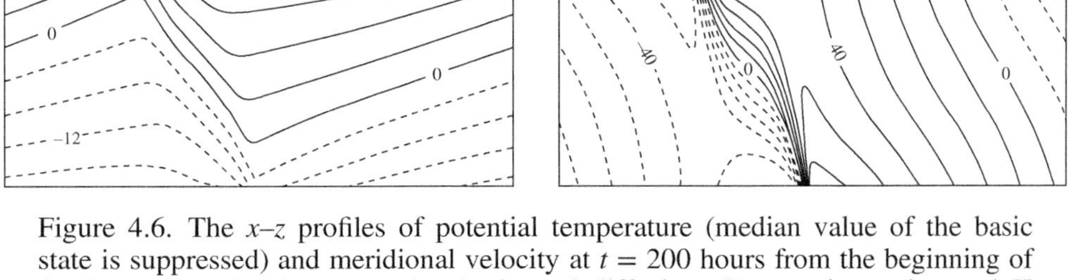

Figure 4.6. The x–z profiles of potential temperature (median value of the basic state is suppressed) and meridional velocity at t = 200 hours from the beginning of the integration for the case using horizontal diffusion. Contour intervals are 4 K and 10 m s^{-1} (excerpted from figure 2 in Nakamura, 1994).

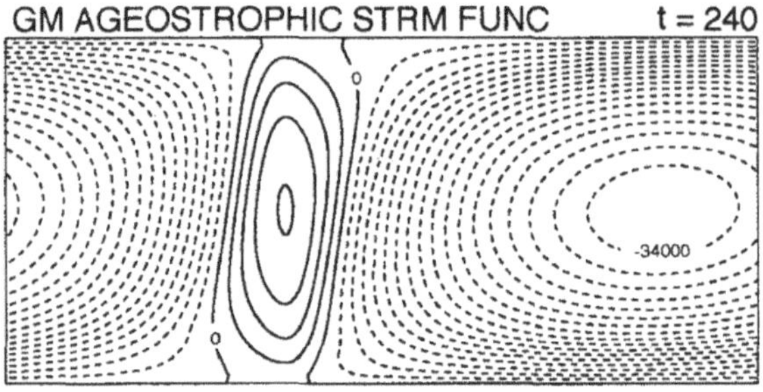

Figure 4.7. Ageostrophic streamfunction at t = 240 hours (excerpted from Nakamura, 1994).

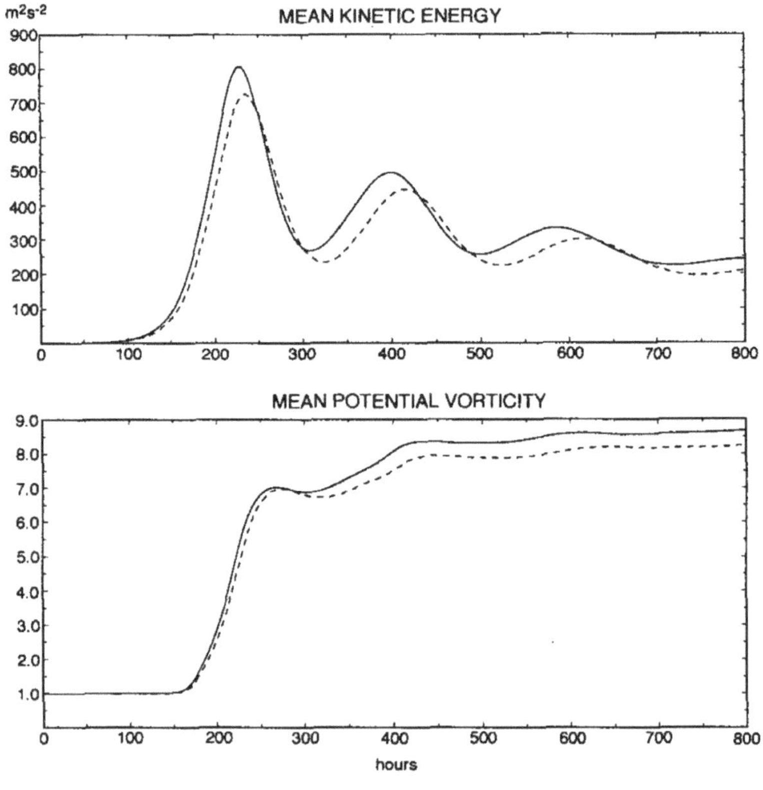

Figure 4.8. Evolution of domain-averaged eddy kinetic energy (top) and potential vorticity (bottom) for the simulation with horizontal diffusion (excerpted from Nakamura, 1994). Dashed curves for the counterpart primitive equations model results reported by Nakamura and Held (1989).

breakdown of the elliptic condition of the SE equation did not occur in that calculation. According to the analytic solution of Hoskins and Bretherton (1972), frontal collapse should occur in such an inviscid model. The conclusion should not be dependent upon which algorithm one chooses to apply in a particular calculation.

4.3 Two-Dimensional SG Frontal e-Charney Model Analysis

The objective of this illustrative analysis is to simulate model fronts that would only develop near the bottom surface. The model setting is a 2D SG model with the reference state that has *an equivalent-Charney basic flow with the beta-effect.* In Section 3.2 we determined the unstable modes of the e-Charney model. We will use a particular unstable mode of this model as the initial disturbance of this analysis. It is recapped in Figure 4.9 for easy reference, although we discussed extensively how it is determined as well as its characteristics in Section 3.2.

We are now ready to perform an analysis of frontal dynamics with this SG e-Charney model. Using the nondimensionalization defined in Section 2.6, we can write the nondimensional SG tendency equation for this model as (dropping the tilde notation):

$$P_t = -(z+u)P_x - wP_z - u_x v_x - w_z \psi_{zz} - v_z(w_x + u_z) - \beta v, \tag{4.1}$$

where $P = \psi_{xx} + \psi_{zz}$. The boundary conditions under consideration are being cyclical in x and

$$\begin{aligned} @\ z=0, \quad & \psi_{zt} = -uv_z + v \\ @\ z=1, \quad & \psi_{zt} = -(z+u)v_z. \end{aligned} \tag{4.2}$$

The corresponding SE equation for this model is:

$$(1+\lambda v_x)\eta_{zz} - 2\Gamma\lambda\theta_x\eta_{xz} + (1+\Gamma\lambda\theta_z)\eta_{xx} = 2\lambda v_x.$$

Boundary conditions: (4.3)

$$\eta(x,0) = 0, \ \eta(x,1) = 0.$$

Although the parameter β does not appear in the SE equation, its impact on the ageostrophic circulation is felt through the geostrophic velocity v and the related potential temperature θ. The same algorithm is applied to determine the numerical solution of this system of equations, as we have done in the case of the SG Eady model.

A surface front has developed to its maximum intensity at the 252nd time-step, corresponding to day 2.6. The frontal development is shown in Figure 4.10 in terms of the potential temperature disturbance field. The western boundary of the warm anomaly sector at z = 0 is located at $xk = 5.9$ on day 2.6. That boundary was at $xk = 4.3$ on day 1.5 (not shown) and $xk = 2.3$ at t = 0 (Figure 4.6b). Therefore, the disturbance moves eastward in a speed close to the phase velocity of the initial wave and its movement slows down somewhat as the front strengthens. The disturbance intensifies as it moves eastward. The maximum values of θ on day 0, day 1.5 and day 2.6 are equal to 0.8, 3.3 and 9.4 respectively. The eastward tilt of the cold-to-warm boundary in $\tilde{\theta}$ becomes slightly westward on day 2.5, whereas the warm-to-cold boundary tilts becomes more eastward, similar to what we have seen in the SG Eady model. The width of the warm sector progressively contracts while the width of the cold sector correspondingly expands.

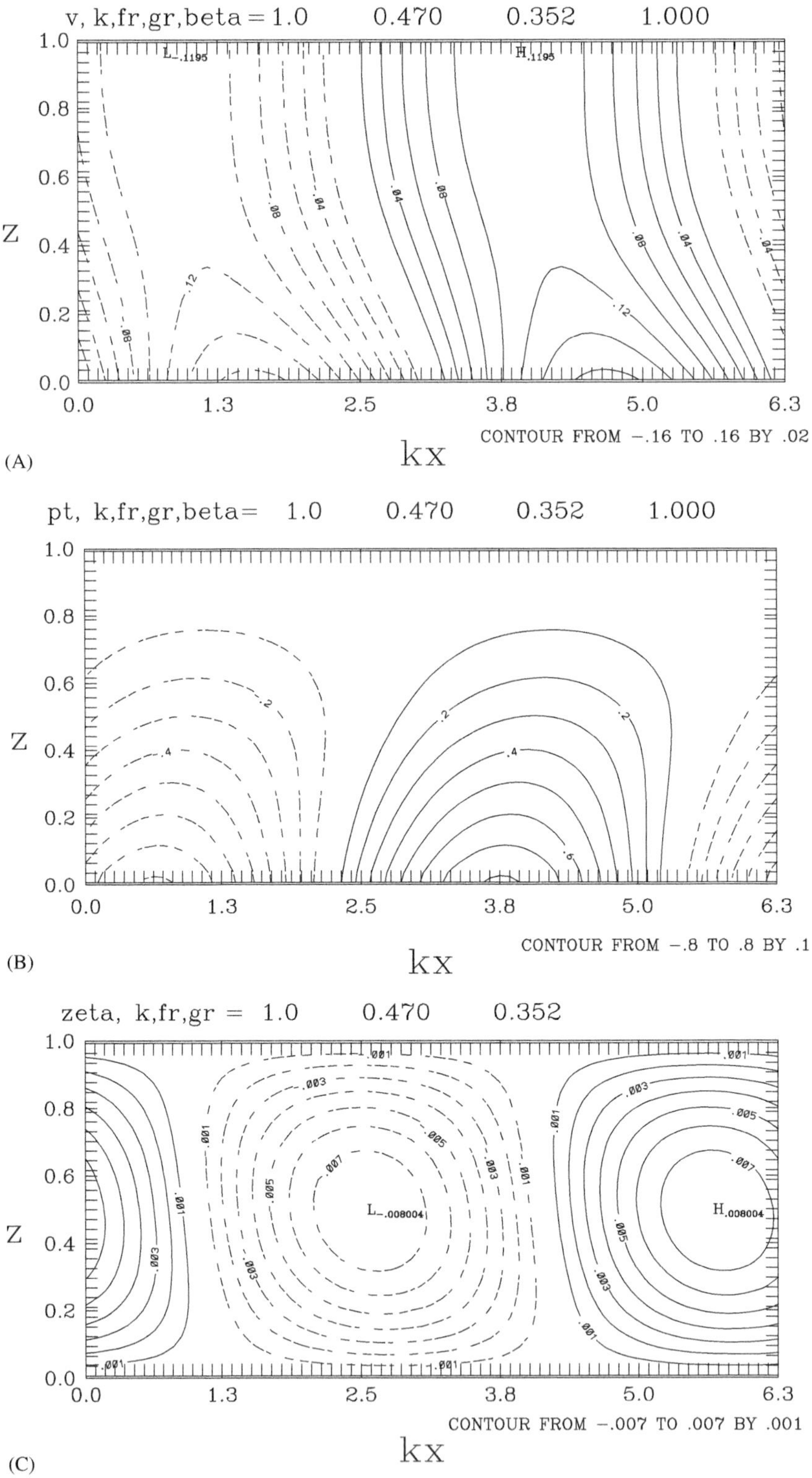

Figure 4.9. Structure of the unstable mode for $(\tilde{\beta}, \tilde{k}) = (1.0, 1.0)$ in an e-Charney model: (a) the geostrophic along-front velocity field in units of $f_o\lambda L/N = 30$ m s^{-1}; (b) the potential temperature field in units of $\Theta = 3$ K; (c) the ageostrophic streamfunction in units of $f_o\lambda DL/N = 3 \times 10^5$ m^2 s^{-1}.

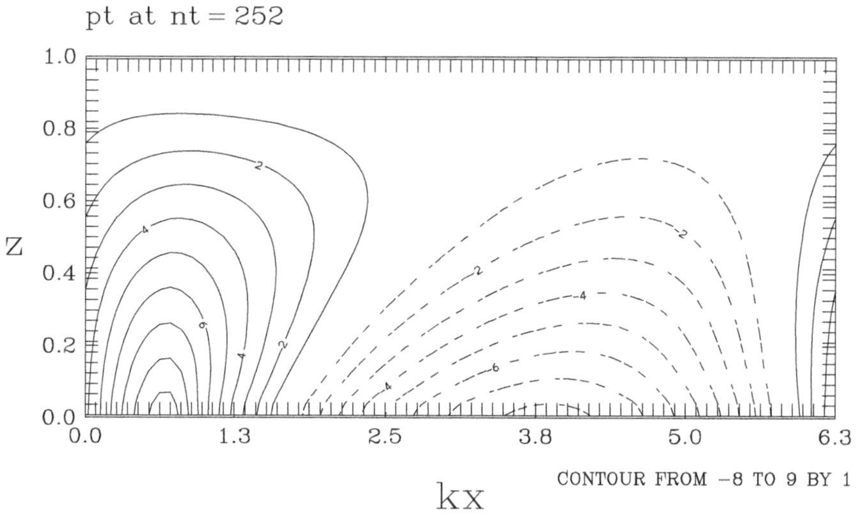

Figure 4.10. The potential temperature field of the disturbance in units of $\Theta = 3\,\mathrm{K}$ on day 2.6.

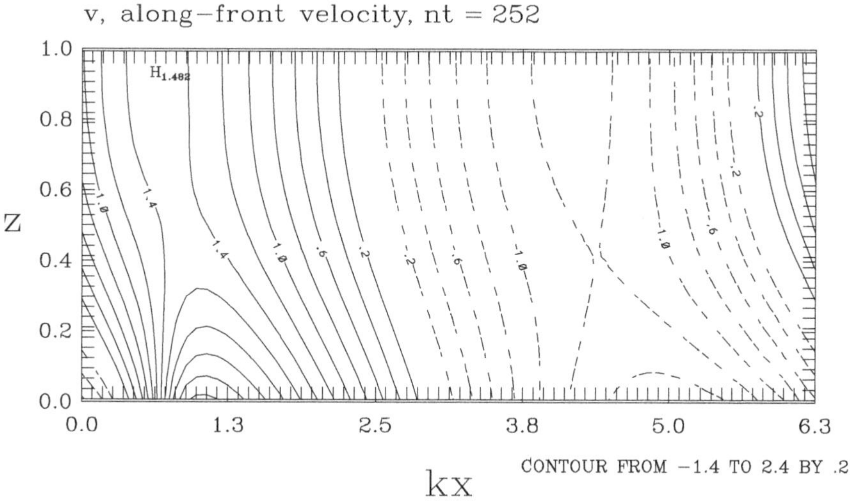

Figure 4.11. Structure of the geostrophic velocity field $v(x, z)$ on day 2.6 in units of $\lambda D = 30\,\mathrm{m\,s^{-1}}$.

The structure of the geostrophic velocity field is shown in Figure 4.11. The southerly (northerly) flow is strong where the temperature is warm (cold). The width of the southerly flow sector is considerably narrower, whereas the width of the northerly flow sector is much wider. It follows that the southerly flow is much stronger than the northerly flow as mass conservation would require.

Figure 4.12 shows the corresponding distribution of the ageostrophic circulation. Panel (a) shows the ageostrophic horizontal velocity field and panel (b) shows the vertical velocity field. There is highly concentrated ascent ahead of the cold front.

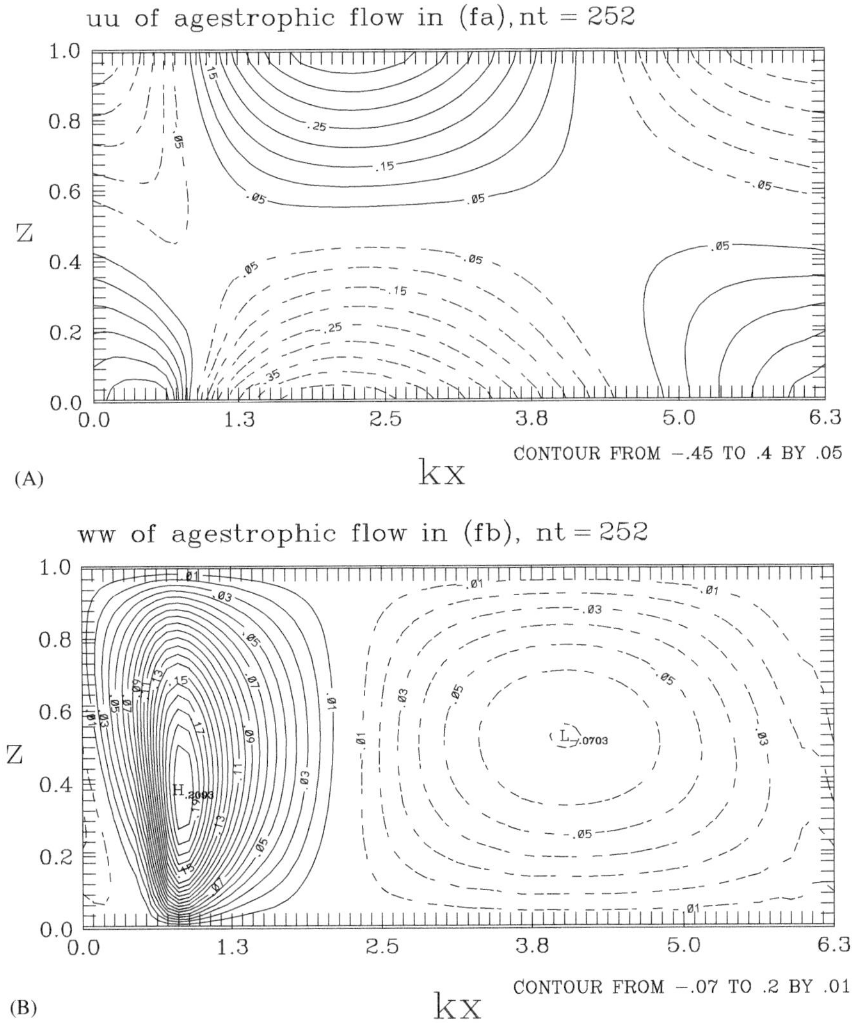

Figure 4.12. Structure of (a) the ageostrophic velocity $u(x,z)$ in units of $\lambda D = 30\,\mathrm{m\,s^{-1}}$ and (b) vertical velocity $w(x,z)$ in units of $\lambda D f_o/N = 0.3\,\mathrm{m\,s^{-1}}$ on day 2.6.

The maximum ascent is located at $(xk, z) = (0.7, 0.4)$ over the warm sector. The maximum value is equal to 0.21 corresponding to 6 cm $\mathrm{s^{-1}}$ arising from the strong convergence ahead of the front near the surface boundary and weaker divergence near the top boundary. There is no tilt in this field. The descent is much broader and weaker, mostly in the cold sector. This is therefore a thermally direct cell about the cold front favorable for intensification of the disturbance as a whole. The overall structure of the flow associated with a surface front is better simulated in this 2D SG e-Charney model than the 2D SG Eady model.

The elliptic condition for the SE equation breaks down and the integration stops shortly after day 2.6. We show the distribution of the absolute vorticity and total

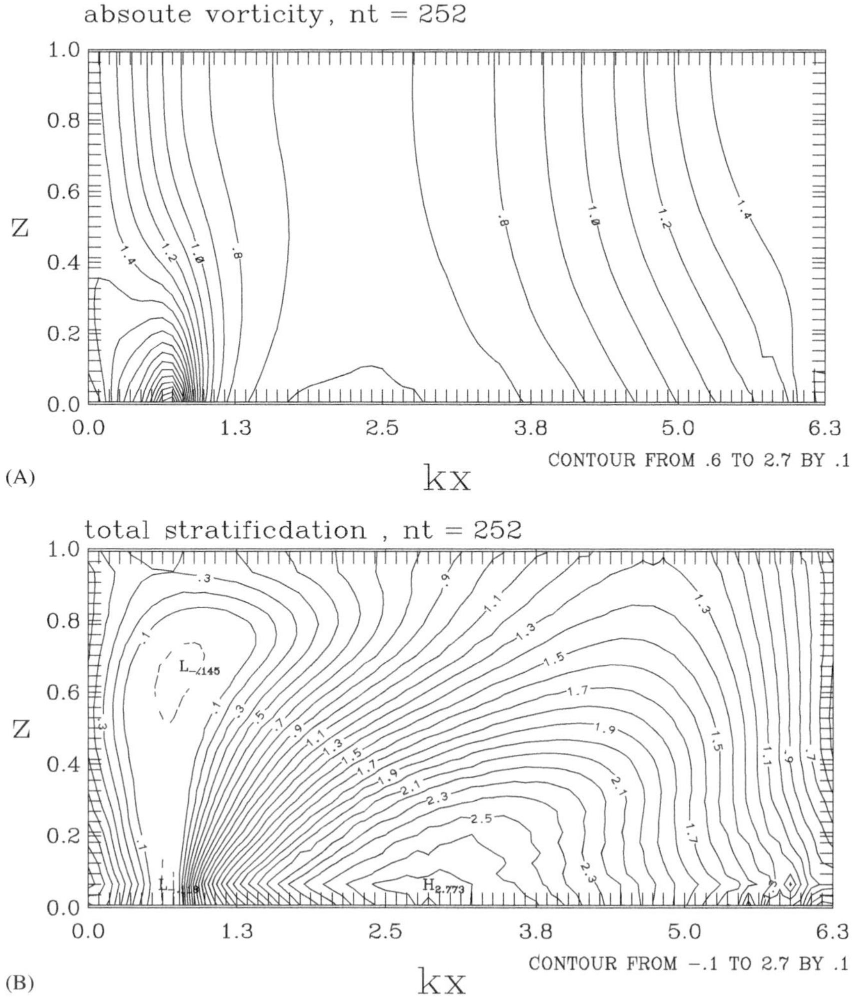

Figure 4.13. Distribution of (a) the absolute vorticity $(1+\lambda v_x)$ in units of $f_o = 10^{-4}\,\mathrm{s}^{-1}$; and (b) the total stratification $(1+\lambda\Gamma\theta_z)$ in units of $N^2 = 10^{-4}\,\mathrm{s}^{-2}$ at the 252nd time-step (day 2.6).

stratification on day 2.6 in Figure 4.13. Panel (a) shows that the absolute vorticity at the cold front is quite strong at this time, reaching a value of $\approx 2.7 f_o$. Note that the anticyclonic vorticity is much smaller than the cyclonic vorticity in this flow. So much so that the absolute vorticity is positive everywhere, implying that this flow remains inertially stable. A deep layer of almost neutrally buoyant air develops slightly ahead of the cold front in the warm sector. The breakdown of the elliptic condition for the SE equation is clearly attributable to the stratification becoming negative ahead and over the cold front. In other words, the front has reached the maximum intensity within the SG dynamic framework. It again suggests that frontal collapse is suppressed by small-/meso-scale instability rather than by frictional damping in the atmosphere.

Table 4.1. Properties of the three quasi-geostrophic models of cyclogenesis

Model properties	Eady model	Charney model of finite depth
f	f_o	$f_o+\beta y$
$\overline{u}$	$\lambda z,\ 0\leq z\leq D$	$\lambda z,\ \ 0\leq z<D$
		$\overline{u}_z=0,\ \ D-\delta<z\leq D$
		$\lim \delta\rightarrow 0$
S	N^2	N^2
$\overline{q}_y$	0	β
q'	$\psi'_{xx}+\frac{f_o^2}{S}\psi'_{zz}$	$\psi'_{xx}+\frac{f_o^2}{S}\psi'_{zz}$

4.4 Limitations of 2D Frontal Models

Although much can be learned about frontal dynamics with the use of a 2D SG model, it has significant limitations. Apart from the fact that fronts in the atmosphere generally have pronounced curvature, there is a deeper reason as to why a 2D model is overly restrictive. In the case of a 2D front, the roles of the geostrophic and ageostrophic velocity components are entirely separable. The contraction of the cross-front scale only stems from the deformation effect of the ageostrophic velocity component. The increase of cross-front thermal contrast solely arises from cold advection to the west and warm advection to the east by the geostrophic velocity component alone. Such separation of the roles is the reason that a 2D SG model cannot dynamically distinguish a cold front from a warm front. It is more suitable for capturing the dynamic processes in a cold front than in a warm front. An illustrative analysis with a 3D SG model presented in Chapter 5 will show that the thermal advection by the geostrophic and ageostrophic components of the flow reinforce one another in a cold front region, but partially counteract one another in a warm front region. That is why cold fronts generally are distinctly stronger than warm fronts.

The mathematical expressions for the basic states and perturbation potential vorticity of the three models under consideration are summarized for easy reference in Table 4.1.

Recap

- Two aspects of frontal dynamics are delineated in two illustrative analyses with two versions of a 2D SG frontal model.
- Two-dimensional SG frontal Eady model: a minimal model analysis using the Eady flow as the reference state and its most unstable mode as an initial disturbance:
  - Fronts develop at the bounding surfaces as the disturbance field sufficiently intensifies and moves eastward.
  - Advective effect of the ageostrophic velocity component on the geostrophic velocity component generates significant higher harmonics in a few days.
  - Superposition of all those harmonics gives rise to pronounced scale contraction and counterpart scale stretching (due to shorter wave components moving slower) at the two bounding surfaces, leading to sharp local gradient of all properties.
  - Deformation of the ageostrophic velocity component tightens surface temperature gradient; inducing ascending motion along a slanting frontal surface.
  - Inexorable development leads to eventual breakdown of the Sawyer–Eliassen equation and hence a frontal collapse.
  - Frontal collapse would not occur when diffusion is incorporated. Instead the flow would undergo oscillations and eventually be equilibrated to a steady state with virtually no lasting front.
- Two-dimensional SG frontal equivalent-Charney model: an analysis using a Charney-like basic flow as the reference state and a corresponding unstable mode as an initial disturbance:
  - Mature surface front emerges in a few days, with pronounced ascending motion ahead of the front.
  - Frontal collapse occurs in about three days as the Sawyer–Eliassen equation breaks down.
- Limitations of 2D models for frontal simulation stems from the fact that the geostrophic and ageostrophic velocity components have distinct and separable roles, making the cold front and warm front dynamically indistinguishable.

5

Three-Dimensional SG Frontal Dynamics

5.1 Introductory Remarks

Building upon what we have learned from the illustrative analyses of 2D frontal dynamics in the previous chapter, we now proceed to analyze 3D frontal dynamics in the context of a semi-geostrophic (SG) model. This generalization is necessary because, as noted earlier, a 2D SG model for surface fronts is too restrictive in that it does not dynamically distinguish between cold and warm fronts. The intrinsic differences between cold fronts and warm fronts can only be simultaneously examined in a 3D setting. That is also the focus of an analysis in Chapter 7 with an even more general model. The analysis here is performed in physical coordinates (x, y, z) rather than in geostrophic coordinates as first reported by Mak (2014).

5.2 Three-Dimensional SG Frontal Model Analysis

The goal of this illustrative analysis is to demonstrate that a 3D SG model would be able to capture the intrinsic dynamics of the life-cycle of fronts in a generic setting compatible with the extratropical atmosphere. The governing equations of this model have been presented in Section 2.4. Let us first outline the steps that need to be taken. We elaborate in Section 2.5.2 on how to calculate the change in the geostrophic streamfunction at each time-step. The method to determine the corresponding change in the ageostrophic velocity at each time-step is discussed in Section 2.5.3. It requires taking two further steps, elaborated in Section 2.5.4. First, we generalize the Sawyer–Eliassen (SE) equation to two coupled partial differential equations that diagnostically relate two ageostrophic potentials to the geostrophic streamfunction. Then we obtain an analytic formal solution of the two ageostrophic potentials that are in the form of two separate functionals of the geostrophic streamfunction.

We prescribe a specific reference state and a correspondingly suitable initial disturbance. The properties of the reference state are $\overline{u}(y, z), \overline{\varphi}(y, z), \overline{\theta}(y, z), \theta_o(z)$. We construct such a state compatible with the observed structure of the winter and

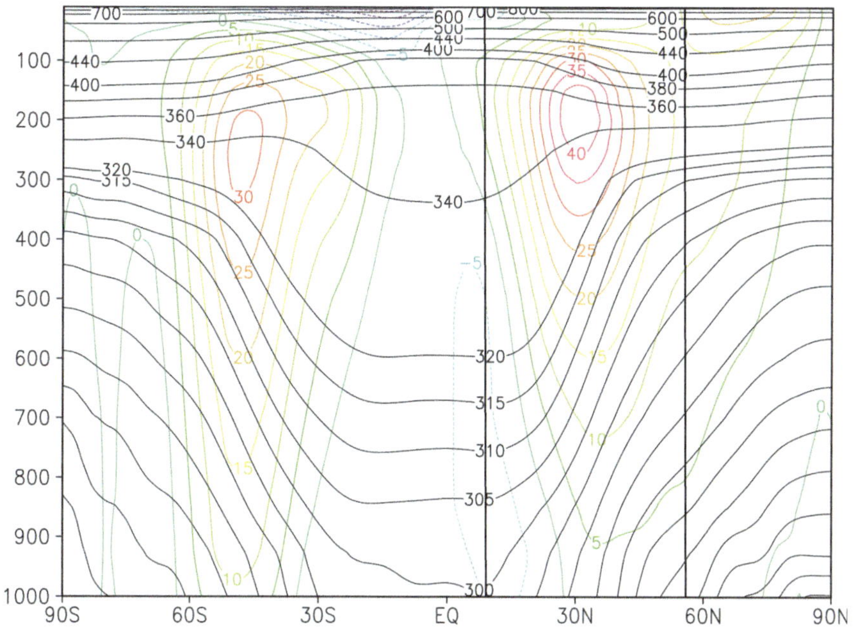

Figure 5.1. Observed zonal and winter mean zonal velocity (color contours) and potential temperature (black and white contours), taken from Mak (2011).

zonal mean state of the atmosphere shown in Figure 5.1. In the mid-latitudes of each hemisphere, there is a zonal baroclinic jet centering at about 30–35° latitude and at the 200 mb level. The baroclinic jet in the winter hemisphere is somewhat stronger than in the southern hemisphere. The latitude belt in the northern hemisphere is highlighted by the two vertical lines.

5.2.1 *Reference State*

The zonal flow of the reference state $\overline{u}(y,z)$ should have a structure somewhat similar to the observed jet. It is to be in hydrostatic and geostrophic balance, and thus should satisfy the requirement of thermal wind balance leading to corresponding choices for $\overline{\phi}(y,z)$ and $\overline{\theta}(y,z)$. The reference state also should have a stable stratification everywhere, $\theta_o(z)$. Such would be an environment in which atmospheric fronts in mid-latitude are expected to develop. It would therefore be meaningful to use a simple reference state for an investigation of frontal dynamics in the extratropics in the following analytic form. It consists of a baroclinic Gaussian jet in a domain $0 \leq z \leq D, 0 \leq y \leq Y$ as:

$$\overline{u}(y,z) = \lambda z \exp(-((y-c)/b)^2). \tag{5.1}$$

The parameters λ, c, b would specify the strength of the baroclinic shear, the latitudinal location of the jet core and the width of the jet, respectively. Since $\overline{u}$ is to

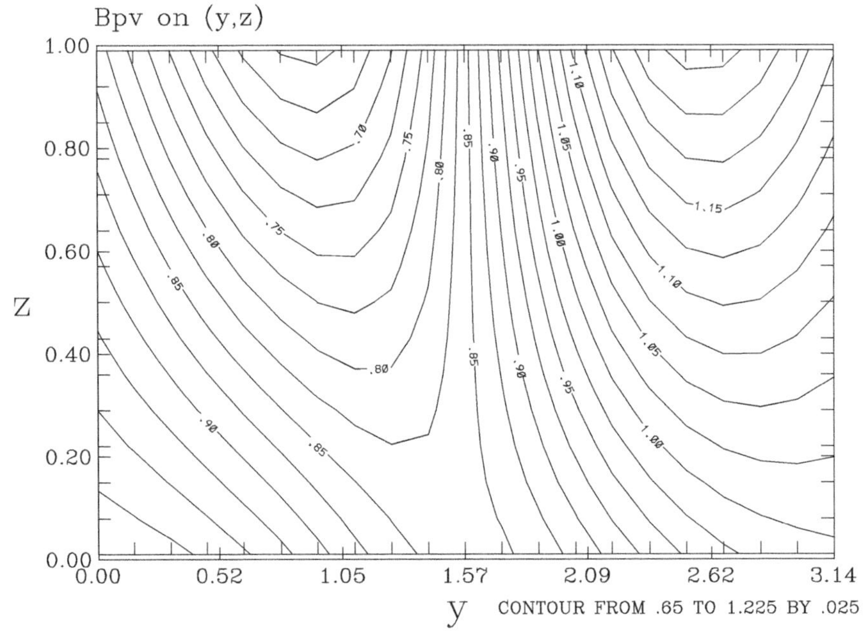

Figure 5.2. Distribution of the nondimensional potential vorticity of the background state $\tilde{\overline{q}}$ in units of $\frac{\theta_{oo}N^2 f}{g}$ (taken from Mak, 2014).

be in thermal wind balance with the potential temperature of the reference state, the latter must have the form of

$$\overline{\theta}=\frac{f\theta_{oo}\lambda b\sqrt{\pi}}{2g}\left\{erf\left(-\frac{c}{b}\right)-erf\left(\frac{y-c}{b}\right)\right\}, \tag{5.2}$$

where $erf(\eta)=\frac{2}{\sqrt{\pi}}\int_0^\eta e^{-t^2}dt$ is the error function. In addition, there is a constant stratification associated with a constant Brunt–Vaisala frequency $N=\sqrt{\frac{g}{\theta_{oo}}\frac{d\theta_o}{dz}}$.

The definition of SG potential vorticity q is a conservative property of this model, given in (2.8). The potential vorticity of the reference state is:

$$\overline{q}=\frac{\theta_{oo}}{g}(-f(\overline{u}_z)^2+(f-\overline{u}_y)N^2). \tag{5.3}$$

All equations are nondimensionalized according to Appendix A. The distribution $\overline{q}$ in nondimensional form is shown in Figure 5.2. The change of sign in $\partial\overline{q}/\partial y$ in the domain as seen in Figure 5.2 assures that the necessary condition for the instability of a quasi-geostrophic (QG) disturbance is satisfied.

5.2.2 Initial Disturbance

The next preliminary task is to prescribe an initial disturbance that could be expected, starting from an appropriate initial state, to intensify in time. The model

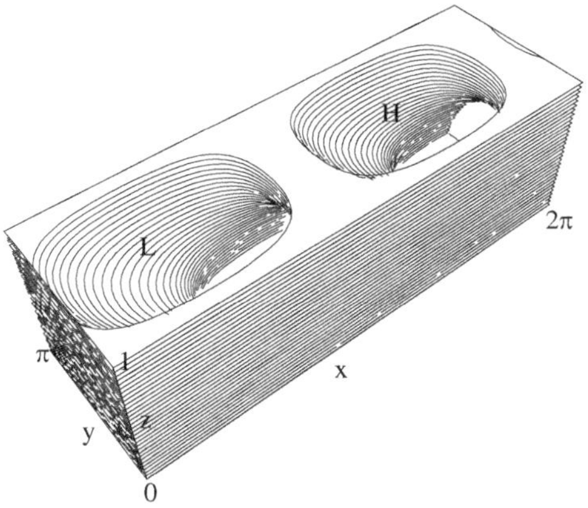

Figure 5.3. A 3D portrait of the distribution of the nondimensional ψ field of an unstable Eady wave for $k=l=1$.

has three effective dependent variables: one geostrophic streamfunction ψ and two ageostrophic potentials η and χ. They are prescribed on the basis of the following considerations. The basic flow in the Eady model of baroclinic instability may be viewed as the limiting case of the baroclinic jet under consideration: an infinitely wide baroclinic jet. For convenience, we therefore use the analytically known geostrophic streamfunction of a 3D unstable Eady wave in a channel domain as the initial ψ field. Its 3D structure is shown in Figure 5.3. It has a westward vertical tilt and a sinusoidal horizontal structure in both x- and y-directions. The corresponding η and χ fields, and hence the ageostrophic velocity $(\tilde{u}, \tilde{v}, \tilde{w})$ field, are computed with (2.15) and (2.16) associated with the initial ψ field above. Such an initial disturbance is self-consistent and may be expected to intensify immediately. Its amplitude is set such that the maximum geostrophic speed is equal to 4 m s^{-1}.

The nondimensional length of the domain is $kx_{\text{max}}=2\pi$, and the width of the domain is $\ell Y=\pi$, where (k,ℓ) are the horizontal wavenumbers of the initial wave. We particularly consider a square wave, $(k,\ell)=(1,1)$. The total potential temperature field at the bottom boundary $z=0$ is plotted and superimposed onto the geostrophic velocity field of the initial disturbance. Since we have $\overline{u}(y,0)=0$, the vector field in Figure 5.4a is also the total geostrophic velocity field. This velocity field varies sinusoidally in both the x- and y-directions. There is a cyclonic flow around a low-pressure center near $(kx,\ell y)=(1.9\pi,0.5\pi)$. There is an anticyclonic flow around a high-pressure center near $(kx,\ell y)=(0.9\pi,0.5\pi)$. The contours indicate the isotherms. The temperature at the southern and northern boundaries is 300 K and 270 K respectively. This geostrophic flow component gives rise to maximum warm (cold) advection at a quarter of a wavelength to the east (west) of the low-pressure center. The thermal advection is largest at the mid-width of the channel domain, $\ell y=0.5\pi$.

The ageostrophic circulation associated with this geostrophic streamfunction is computed using the formal solutions (2.15) and (2.16). The horizontal ageostrophic

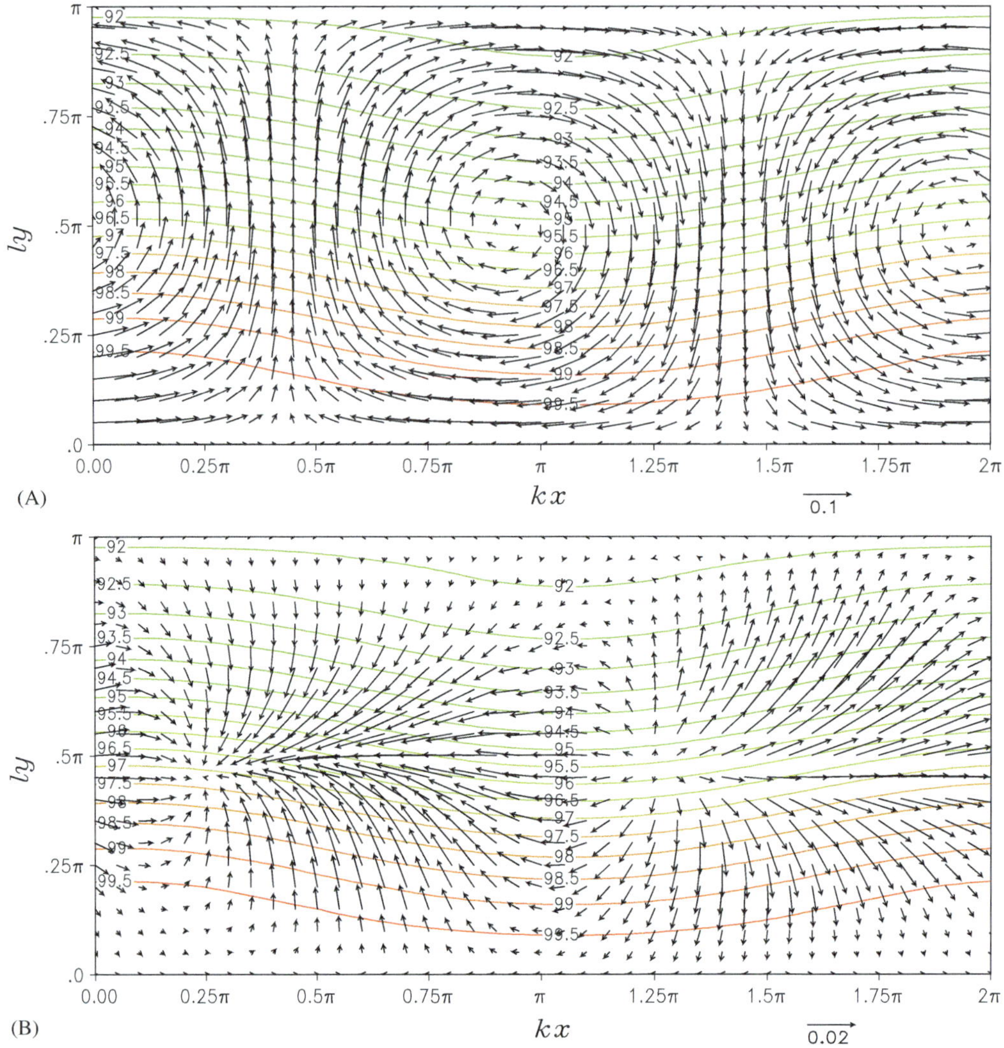

Figure 5.4. Structure of the initial state at the bottom surface $z=0$: (a) Geostrophic velocity $(\overline{u}+\hat{u}, \hat{v})$ vector field in units of $\lambda D = 40$ m s^{-1}; and (b) ageostrophic velocity $(\tilde{u}, \tilde{v})$ vector field together with potential temperature $(\overline{\theta}+\hat{\theta})$ field (values decreasing from red to green) in units of $\Theta = 3$ K (taken from Mak, 2014).

velocity field at each level, exemplified by that at $z=0$, has the appearance of a magnetic field with a center of convergence and a center of divergence resembling the two poles of a bar-magnet, albeit not exactly symmetrical between them. We may anticipate such a signature as a broad characteristic a priori because the ageostrophic velocity vector is, after all, 3D non-divergent, just like a magnetic field vector. At the bottom boundary, the SG $(\hat{u}, \hat{v})$ vectors converge toward a point at the half-width of the domain from all directions,$(kx, \ell y) = (0.25\pi, 0.5\pi)$, and also diverge to all directions from another point at the half-width, $(kx, \ell y) = (1.25\pi, 0.5\pi)$ (Figure 5.4b). The ageostrophic velocity is strongest at the latitude of the center of the baroclinic jet in the reference state, $ly = 0.5\pi$. The nonvanishing meridional component of this perturbation is primarily a consequence of the fact that the jet of the zonal flow in the reference state has a jet structure. The ageostrophic velocity component is about one-fifth as strong as the geostrophic velocity component. In light of Figure 5.4a and 5.4b, it is worth emphasizing that there is not a trace of front in the initial disturbance.

The pattern of thermal advection by this ageostrophic velocity is different from that by the corresponding geostrophic velocity. Specifically, the geostrophic velocity gives rise to cold (warm) advection in the quarter-wavelength to the west (east) of the low-pressure center at $z = 0$. In contrast, the ageostrophic velocity gives rise to cold (warm) advection in the SW and NE (SE and NW) quadrants relative to the point of convergence. The latter itself is located at a quarter-wavelength to the east of the low-pressure center. In other words, the cold advections due to the two velocity components largely reinforce one another in the region $(\pi < kx < 2\pi,\ 0 < \ell y < 0.5\pi)$, as does the warm advections in the region $(0 < kx < \pi,\ 0 < \ell y < 0.5\pi)$. But the thermal advections by the two velocity components oppose to one another in the northern half of the domain, $0.5\pi < \ell y < \pi$. The combined influence of such initial thermal advection has the potential of inducing two different fronts: a cold (warm) front to the south-southwest (east-northeast) of the low-pressure center. This is a unique feature of 3D SG dynamics, very pertinent to how cold and warm fronts subsequently evolve in different manners, as we observe in the atmosphere.

The vertical structure of this initial disturbance is not shown for the sake of brevity, but can be readily visualized. Since it is a baroclinic unstable wave, the vector fields of both the geostrophic and ageostrophic velocities have a westward tilt with height. The vertical tilt of the geostrophic velocity is somewhat greater. There is a branch of ascending motion extending upward from the point of convergence at $z = 0$ toward the point of divergence at $z = 1$. There is a compensating branch of descending motion that connects the point of convergence at $z = 1$ to the point of divergence at $z = 0$.

It should be noted that although this initial disturbance is self-consistent in that its geostrophic properties $(\hat{u}, \hat{v}, \hat{\theta})$ and ageostrophic circulation $(\tilde{u}, \tilde{v}, \tilde{w})$ satisfy the generalized form of the SE equation, it is not a normal mode of the reference state. Nevertheless, we may anticipate this initial disturbance to intensify in time and to propagate eastward. It is expected to initially evolve toward the structure of an unstable normal mode. Under the influence of the thermal advection by the two velocity components discussed above, a cold front and a warm front would simultaneously develop in due course. It would, however, be premature to make specific inferences about the frontal development on the basis of the initial tendency. The advective nonlinear dynamics would soon break the initial symmetry of the flow as a result of the spontaneous excitation of the higher harmonics and would also greatly alter the vertical structure of the disturbance (the departure state). The subsequent evolution of the structure can be quite complicated and can only be ascertained by numerical integration.

5.2.3 Algorithm

As elaborated in Sections 2.4 and 2.5, three unknowns describe the state of the flow. Each spatial derivative in the governing equations is replaced with center-difference approximation. At the grid points the unknowns are denoted as $\psi_{i,j,k}$, $\eta_{i,j,k}$ and $\chi_{i,j,k}$, where $(i=1,2,...\,I)$, $(j=1,2,...\,J)$, $(k=1,2,...\,K)$. The grid point values of $G\{\psi\}$ and $H\{\psi\}$ are likewise denoted as $G_{i,j,k}$ and $H_{i,j,k}$. They are arranged as vectors in a specific order: $\boldsymbol{\psi}$, $\boldsymbol{\eta}$, $\boldsymbol{\chi}$, G, and H. For example, $\boldsymbol{\chi} = \{\chi_{i,j,k}\}$ is defined with i systematically varying most rapidly and k most slowly. We apply the following algorithm in the integration. With an initial state, we begin by computing P_t with the RHS of (2.19) as well as Ψ_{zt} at $z=0,D$ with the RHS of $\psi_{zt} = -(\overline{u}+\hat{u}+\tilde{u})\psi_{xz} + (\hat{v}+\tilde{v})(\overline{u}_z+\hat{u}_z)$. Using a particular numerical scheme, we can calculate the new values of P at all interior grid points and the new boundary values of ψ_z at $z=0,D$ after one time-step δt. That would readily allow us to compute the corresponding new values of $\boldsymbol{\psi}=\{\psi_{i,j,k}\}$.

We next compute the corresponding $\boldsymbol{\eta}$ and $\boldsymbol{\chi}$ as follows. A linear second-order differential operator in finite-difference form is matrix operator. We highlight a matrix by an underline in the following expressions. Multiplication of two matrices is denoted by a dot between them. The sequential order of the operators is understood to be from left to right. The solutions of $\boldsymbol{\eta}$ and $\boldsymbol{\chi}$ then take on the following form according to (2.15) and (2.16):

$$\begin{aligned}\boldsymbol{\eta} &= \underline{S_1^{-1}}\cdot\mathbf{G} - \underline{S_1^{-1}}\cdot\underline{R_1}\cdot(\underline{S_2}-\underline{R_2}\cdot\underline{S_1^{-1}}\cdot\underline{R_1})^{-1}\cdot(\mathbf{H}-\underline{R_2}\cdot\underline{S_1^{-1}}\cdot\mathbf{G})\\ \boldsymbol{\chi} &= (\underline{S_2}-\underline{R_2}\cdot\underline{S_1^{-1}}\cdot\underline{R_1})^{-1}\cdot(\mathbf{H}-\underline{R_2}\cdot\underline{S_1^{-1}}\cdot\mathbf{G}).\end{aligned} \tag{5.4a,b}$$

After such computation, we would proceed to integration for the next time-step. The whole procedure can be repeated as long as a balanced flow of the model exists in principle. The computational task is quite time-consuming because we have to evaluate the ageostrophic circulation at each time-step. To speed up the computation, we use a leap-frog scheme to do the time integration. The time-step is small enough that this is not a numerical issue.

5.3 Development of the Model Fronts

Limited by the available computing resources, we only use $I = 41$ grids in the x-direction, $J = 21$ grids in the y-direction and $K = 15$ levels in the z-direction for depicting the model domain. Each matrix is $N\times N$, with $N = I\times J\times K = 12\,915$ being still quite large. The multiple operations of matrix inversion of this size at each time-step are time-consuming. The dimensional grid distance is $\delta x=\delta y\approx 80$ km, and $\delta z\approx 700$ m. The time-step is $\delta t = 15$ minutes. We perform the time integration with a leap-frog scheme instead of a predictor-corrector scheme in order to save computation time by a factor of two. It is encouraging to find that the results using a still

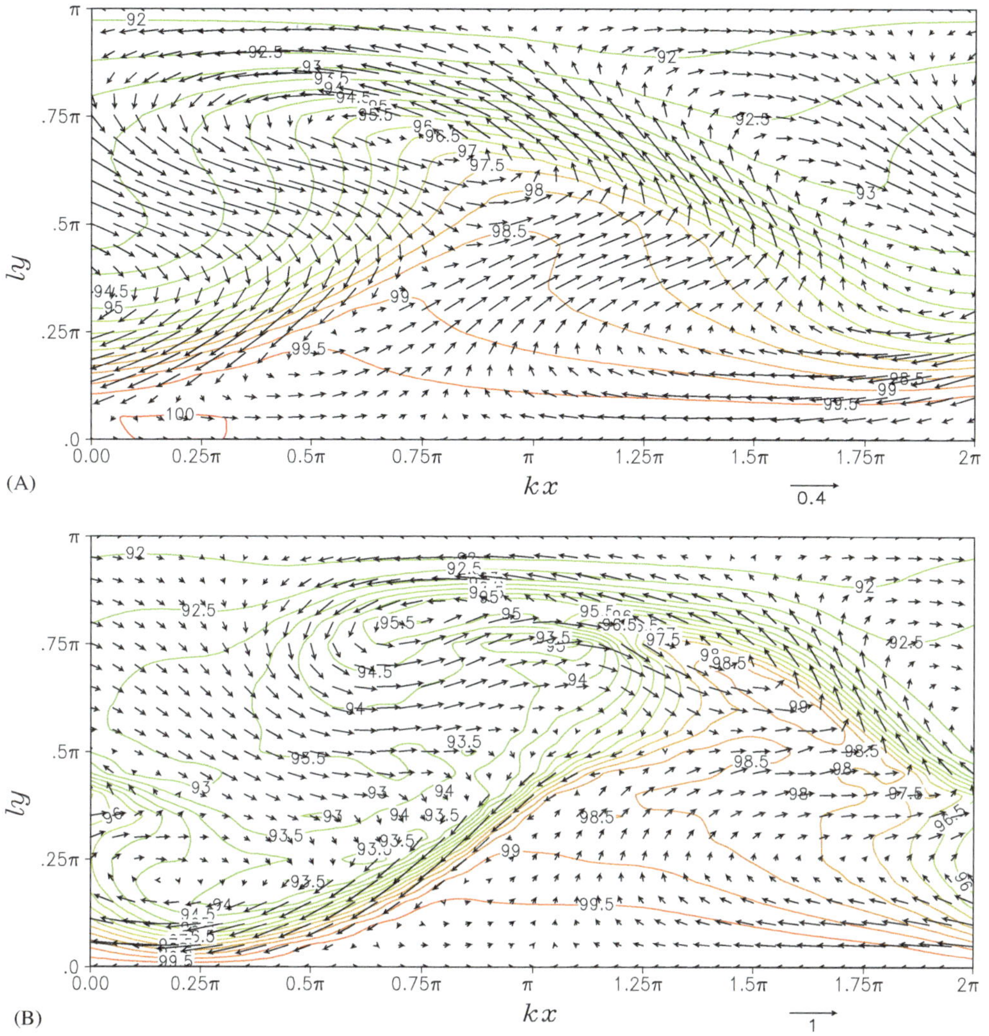

Figure 5.5. Evolution of the geostrophic velocity $(\overline{u}+\hat{u},\hat{v})$ vector field in unit of $\lambda D = 40\,\mathrm{m\,s^{-1}}$ and potential temperature $(\overline{\theta}+\hat{\theta})$ field (values decreasing from red and to green) in unit of $\Theta = 3\,\mathrm{K}$ at the bottom surface $z=0$: (a) on day 3 and (b) on day 5 (taken from Mak, 2014).

lower resolution (I, J, K = 31, 15, 15) are broadly similar. This suggests that the results are robust and trustworthy for the purpose of delineating the frontal dynamics.

The integration is performed with the nondimensional form of the model equations, as given in detail in the Appendix. The integration reveals that the disturbance indeed intensifies rapidly and as a whole moves eastward. Let us begin by examining the evolution of the total geostrophic velocity component $(\overline{u}+u,v)$ and the total potential temperature $(\theta_o+\overline{\theta}+\theta)$. Figures 5.5a and 5.5b show these fields at the bottom boundary on days 3 and 5 respectively. We see that the surface low-pressure center with a cyclonic flow around it intensifies more than the surface high-pressure center with an accompanying anticyclonic flow. The low-pressure center actually moves in the northeast direction, whereas the high-pressure center moves toward the southeast. These broad changes in the disturbance indicate an initial modification of its structure toward an unstable normal mode of the westerly baroclinic jet in the reference state. The distance between the low and high centers is distinctly shorter in one part of the channel and longer in the remaining part. A cold front and a warm front simultaneously develop in the course of cyclogenesis. They become quite pronounced by day 5. These features stem from the nonlinear dynamics particularly attributable to the direct impact of the ageostrophic circulation.

On day 3, the wave disturbance at $z=0$ has intensified fourfold. The surface low-pressure center has displaced northeastward to $(kx, \ell y) = (0.65\pi, 0.75\pi)$ and the surface high-pressure center has displaced southeastward to $(kx, \ell y) = (1.8\pi, 0.35\pi)$. A cold front and a warm front have become well identifiable. The nascent cold front is oriented in the SW–NE direction, located to the southwest of the low-pressure center. The nascent warm front is oriented in WNW–ESE direction located mostly to the east of the low-pressure center. Their combined configuration resembles a capital Greek letter Γ. The disturbance intensifies another 2.5-fold in the next two days. The two fronts are quite pronounced by day 5, transforming to a "T-bone" configuration owing to the fact that the cold front moves eastward faster than the warm front. The curved cold front extends almost to the southern boundary. It seems to tax the adequacy of the grid resolution to its limit. The intensity of the warm front is somewhat weaker. The geostrophic flow is largely but not quite parallel to the cold front and less so to the warm front. It turns sharply and wraps around the low-pressure center, giving rise to greater curvature in the warm front. It is noteworthy that the surface low is distinctly deeper than the surface high and the flow wraps cyclonically around the lows differently to anticyclonically around the lows in this case of non-uniform potential vorticity jet. This is in contrast to what Snyder et al. (1991) found for the case of a uniform potential vorticity jet. In the absence of dissipation, this surface geostrophic flow is strong. This overall surface structure resembles the mature surface fronts simulated with a primitive equation model under comparable conditions.

The evolving ageostrophic velocity field has a more complex structure. It is shown together with the temperature field at the surface boundary on day 3 in Figure 5.6a. The ageostrophic flow has also intensified fourfold. It is mostly perpendicular to the nascent cold front and less so to the nascent warm front. This flow component is convergent in the two frontal regions. By day 5 (Figure 5.6b), it has increased a further twofold in the two frontal regions. It is still more perpendicular to the cold front than to the warm front. The southern boundary has greater impact upon the disturbance than the northern boundary because the cold front extends quite close to the southern boundary.

These fields are comparable to a simulation with an SG model performed in geostrophic coordinates (Snyder et al., 1991), shown in Figure 5.7.

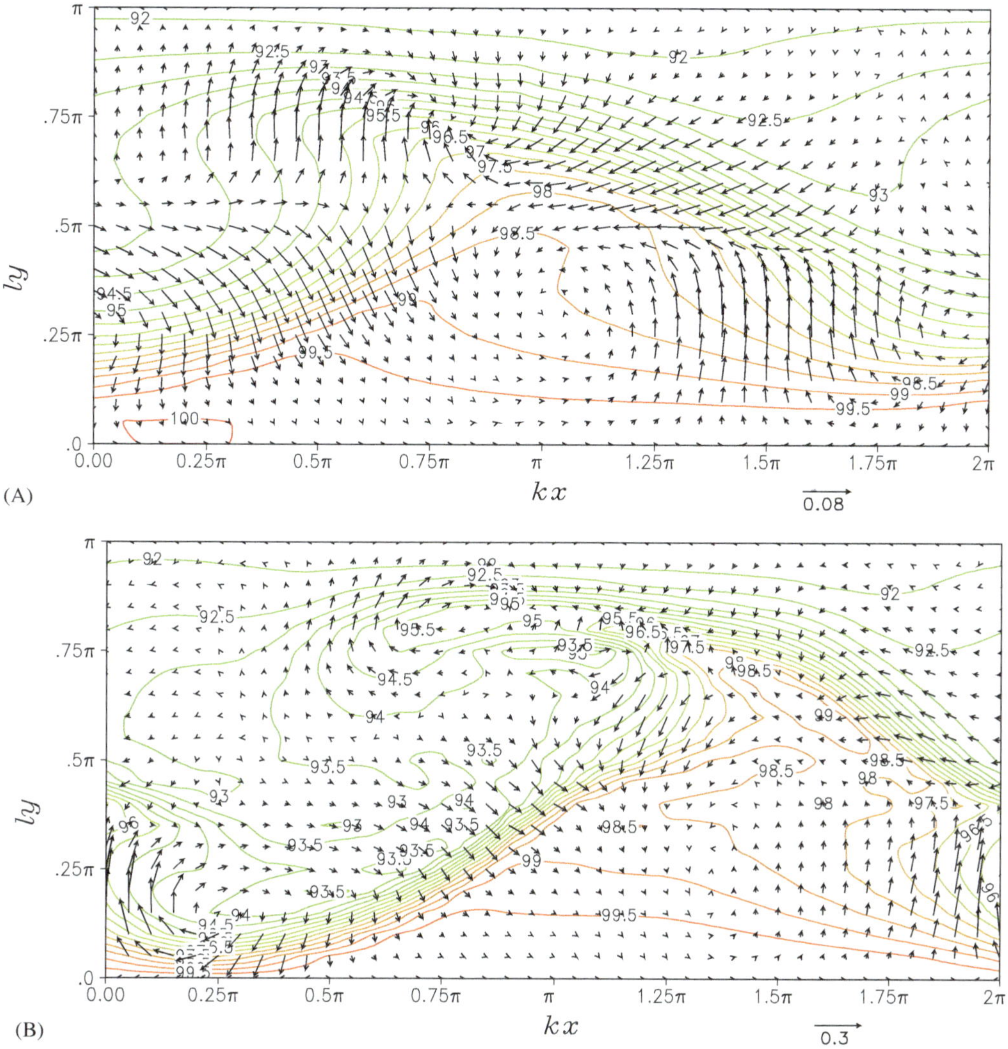

Figure 5.6. Evolution of the ageostrophic velocity $(\tilde{u}, \tilde{v})$ vector field in units of $\lambda D = 40\,\mathrm{m\,s^{-1}}$ and potential temperature $(\overline{\theta} + \hat{\theta})$ field (values decreasing from red and to green) in units of $\Theta = 3\,\mathrm{K}$ at the bottom surface $z = 0$: (a) on day 3; and (b) on day 5 (taken from Mak, 2014).

We next delineate the dynamic roles that the geostrophic velocity and the ageostrophic velocity played in the formation of the fronts from the perspective of thermal advection. Figure 5.8a shows the thermal advection by the geostrophic velocity component, $-(\overline{u} + \hat{u}, \hat{v}) \cdot \nabla(\overline{\theta} + \theta)$, together with the geostrophic velocity field itself at the bottom surface $z = 0$ on day 5. We see strong cold (warm) advection in the cold (warm) front zone. On the other hand, Figure 5.8b shows comparably strong thermal advection by the ageostrophic velocity component, $-(\tilde{u}, \tilde{v}) \cdot \nabla(\overline{\theta} + \theta)$, together with the ageostrophic velocity itself at $z = 0$. Particularly worth emphasizing is that while both the ageostrophic and geostrophic flow components give rise

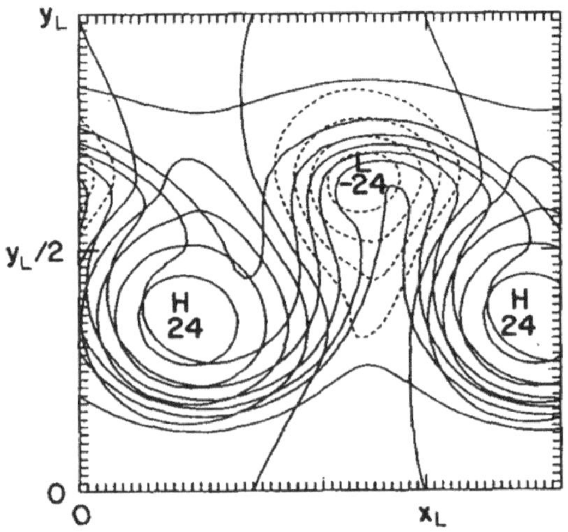

Figure 5.7. Structure of temperature field at the model surface (solid lines 5 K) and perturbation geopotential field (dash lines, CI = 100 m^2 s^{-2}) at day 6.3 (taken from figure 4 of Snyder et al., 1991).

to cold advection in the cold front region, they give rise to opposite thermal advection in the area of warm front. The ageostrophic flow gives rise to moderate cold advection there, whereas the geostrophic flow gives rise to strong warm advection. Consequently, there is a net warm advection in the area of the warm front. This is seen in Figure 5.8c, which shows the thermal advection by the total flow, $-(\overline{u}+\hat{u}+\tilde{u}, \hat{v}+\tilde{v})\cdot\nabla(\overline{\theta}+\theta)$, together with the total velocity field itself.

Associated with the horizontal mass convergence/divergence of the ageostrophic velocity at all levels is a vertical velocity field. To get an overall impression, it suffices to show the vertical velocity field at the mid-level $z=0.5$. It is superimposed onto the ageostrophic velocity at $z=0$ (Figure 5.9). This figure shows the presence of ascending motion slightly ahead of the two frontal regions. The ascending motion ahead of the cold front is stronger than that over the warm front (0.025 versus 0.015). The strong ascent further west of the warm front is located in an area of relatively low temperature. It is therefore thermally indirect, suggesting a process of occlusion there. The strong ascent next to the southern boundary is a direct consequence of the kinematic constraint and should be regarded as an artifact. There is descending motion over two broader areas in support of the divergent surface flow. This general distribution of the ascending and descending motion is compatible with observation. The descending motion is more extensively distributed.

We also compare the ageostrophic velocity and potential temperature fields on day 5 at three levels: z = 0, 0.5, 1 to get a feel for the 3D structure (Figure 5.10a,b,c). The maximum ageostrophic velocity is strongest at the bottom surface, slightly less so at the top boundary and considerably weaker at the mid-level (0.3, 0.2 and 0.06

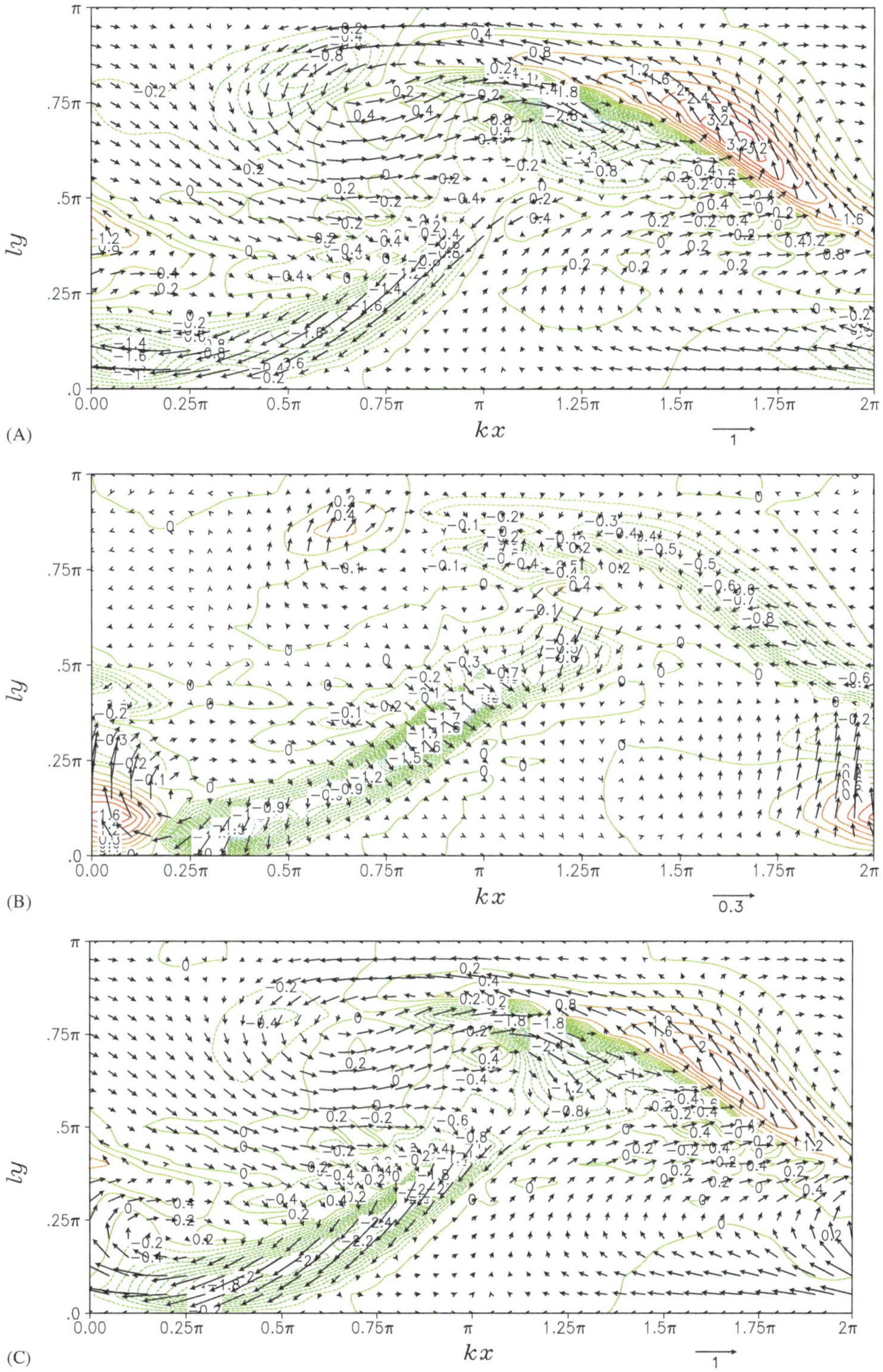

Figure 5.8. Distributions of (a) the geostrophic velocity $(\overline{u}+\hat{u},\hat{v})$ vector field in units of $\lambda D = 40\,\mathrm{m\,s^{-1}}$, with its thermal advection $-(\overline{u}+\hat{u},\hat{v})\cdot\nabla(\overline{\theta}+\hat{\theta})$ (positive values in red and negative values in green); (b) the ageostrophic velocity $(\tilde{u},\tilde{v})$ and its thermal advection $-(\tilde{u},\tilde{v})\cdot\nabla(\overline{\theta}+\hat{\theta})$; and (c) the total velocity $(\overline{u}+\hat{u}+\tilde{u},\ \hat{v}+\tilde{v})$ with its thermal advection $-(\overline{u}+\hat{u}+\tilde{u},\ \hat{v}+\tilde{v})\cdot\nabla(\overline{\theta}+\hat{\theta})$ on day 5 at the bottom surface $z=0$ (taken from Mak, 2014).

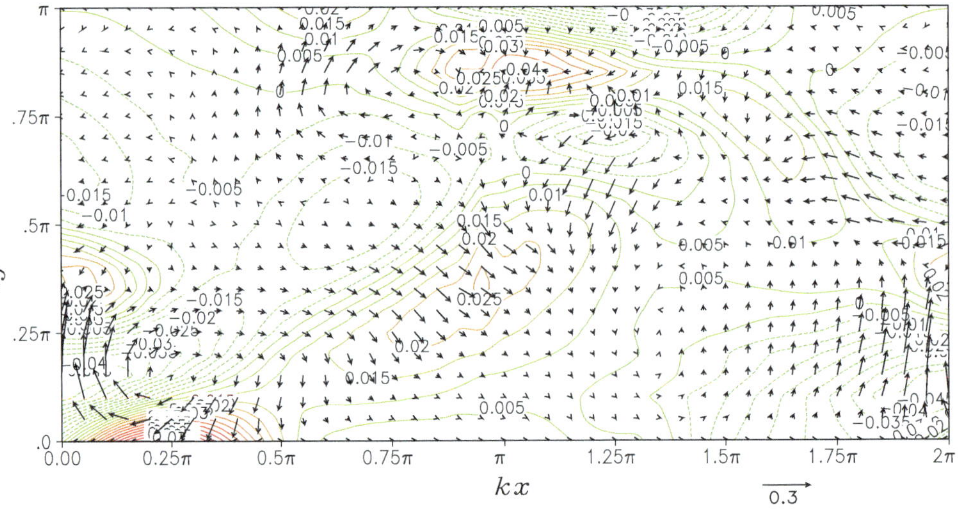

Figure 5.9. Distributions of the ageostrophic velocity vector field $(\tilde{u}, \tilde{v})$ at $z=0$ and the vertical velocity $\tilde{w}$ at $z=0.5$ (positive values in red and negative values in green) and $(\tilde{u}, \tilde{v})$ in units of $\lambda D = 40\,\mathrm{m\,s^{-1}}$ on day 5.

respectively). While the cold front and warm front are most pronounced at the bottom surface, they are somewhat weaker at the top surface. There is no frontal structure to speak of at the mid-level. It is also noteworthy that the weaker front at the top surface located slightly to the west of the surface cold front has a warm-front character, since the ageostrophic velocity is mostly directed from the warm to cold direction. This reveals a cellular structure in the ageostrophic circulation about the cold front.

5.4 Dynamic Nature of Frontal Collapse

Although the solvability condition of the SG equations is satisfied in analytic form, it could break down in a numerical integration when the model grids become inadequate to depict a progressively sharpening front. One encounters this even in a 2D SG front model. According to the analytic solution *in geostrophic coordinates* for the evolution of an initial 2D unstable Eady wave, the model fronts would inexorably develop toward a singularity (Hoskins and Bretherton, 1972). But the singularity is unreachable in a numerical simulation because the grids could not possibly depict the true solution indefinitely. The SE equation in finite-difference form would eventually cease to be elliptic due to accumulative truncation errors in the coefficients. When that happens, a numerical balanced model would have reached its limit of validity. The breakdown of the solvability condition manifests through the appearance of negative potential vorticity in the frontal regions. From a physical point of view, secondary instability should occur in a counterpart primitive equation model when the shear in a frontal region becomes sufficiently strong. For this reason, we may interpret this circumstance as the demise of a 2D SG model front.

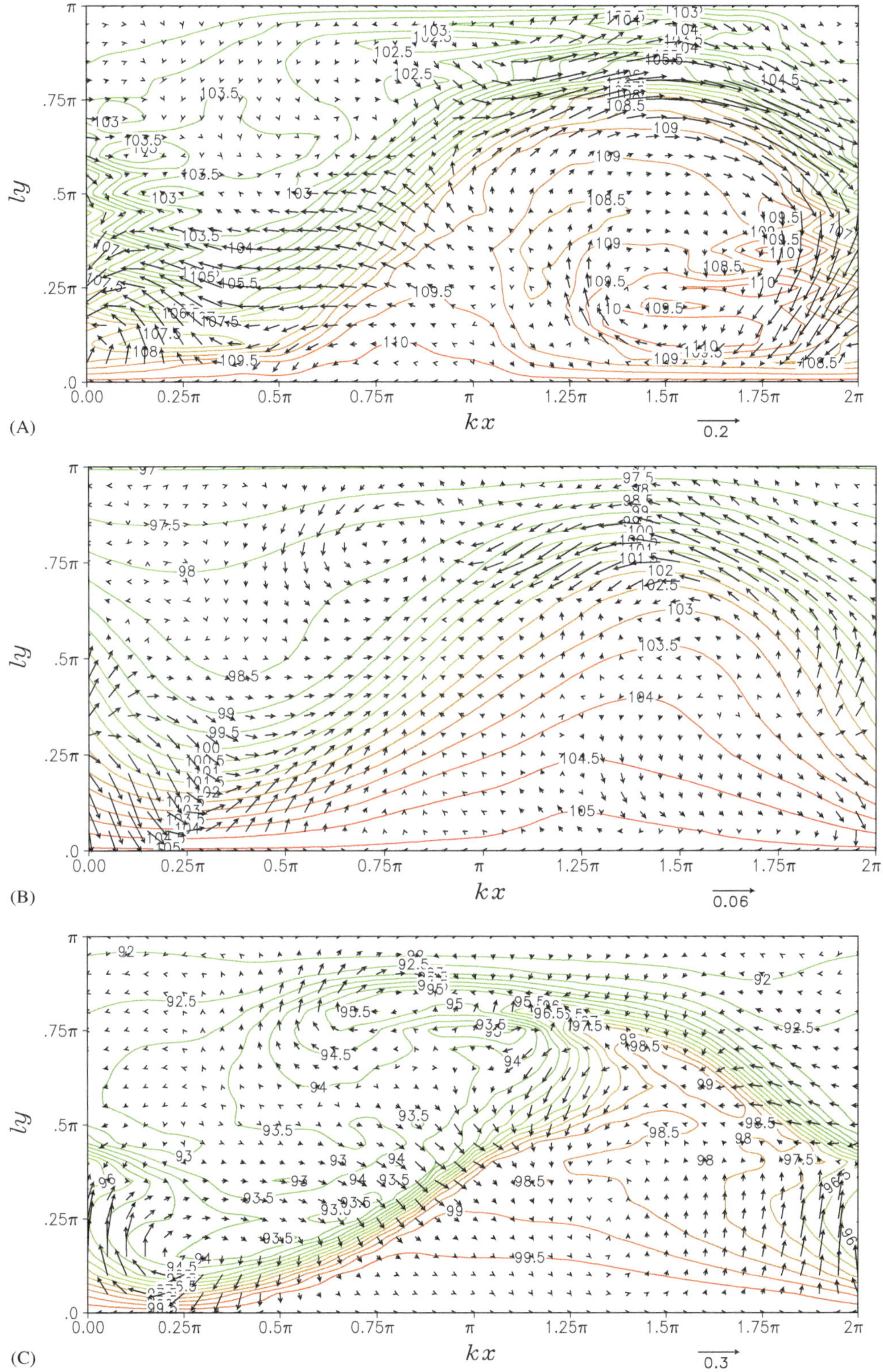

Figure 5.10. Three-dimensional structure of the ageostrophic velocity $(\tilde{u}, \tilde{v})$ and total potential temperature θ_{total} on day 5 shown at (a) $z = 1$, (b) $z = 0.5$, and (c) $z = 0$ (taken from Mak, 2014).

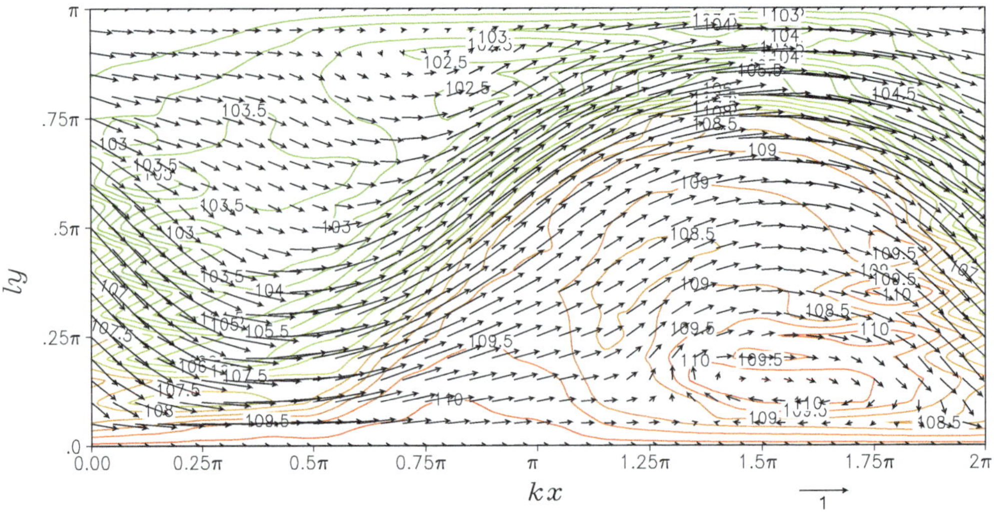

Figure 5.11. Distributions of the total potential temperature (values decreasing from red to green) in units of $\Theta = 3\,\mathrm{K}$ and the total velocity vector field $(\bar{u} + \hat{u} + \tilde{u},\ \hat{v} + \tilde{v})$ at the top boundary $z = 1$ in units of $\lambda D = 40\,\mathrm{m\,s^{-1}}$ on day 5 (taken from Mak, 2014).

The width of our 3D model surface cold front has contracted to about twice the distance between two adjacent grid points on day 5 (Figure 5.5b). In other words, it would soon reach the limit of representation by this resolution, signaling that the front is evolving toward a singularity. It is not surprising to find that the simulation stops shortly after day 5 because computational noise begins to emerge and soon contaminates all fields. Clear evidence of this adverse impact is seen in the upper-level temperature field on day 5. Figure 5.11 shows the total velocity and potential temperature fields at the top boundary $z = 1$ on day 5. It has the appearance of an open wave made up of a strong wave field and a strong zonal wind at the jet core. The flow is sub-geostrophic in the trough region and super-geostrophic in the ridge region, as we expect. Although the velocity field is relatively smooth, the temperature field (proportional to ψ_z) is noisy on the western side of the trough. The local zig-zag feature in the temperature field is an ominous sign of emerging computational noise. The wave is so strong at this time that the temperature distribution resembles a swell of sea breaking upon the shore.

To further delineate the underlying reason for the noise, we examine the two parts of the potential vorticity, q_H and q_V, at $z = 0$ on day 5 with (5.A4) and (5.A5). These are superimposed onto the geostrophic velocity field in Figure 5.12a,b. We see that while such values of q_V are positive everywhere, q_H has negative values in the two frontal bands. The latter stems from the large horizontal gradient of θ in the frontal zones, which obviously cannot be resolved well by the grids. The sum of the two parts has net negative values in the frontal zone. The appearance of local negative values of q implies that the condition of solvability for the 3D generalized SE

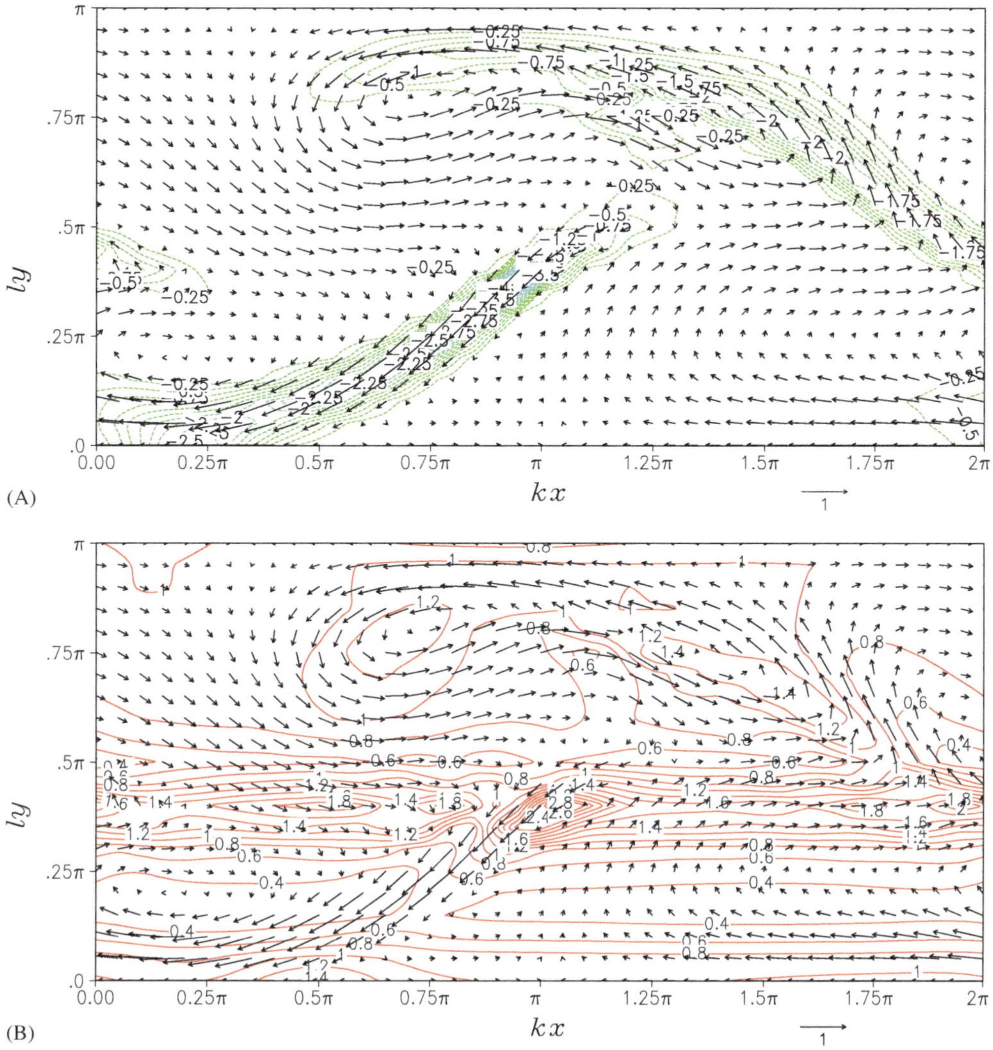

Figure 5.12. Distribution of the two parts of SG potential vorticity of the flow (a) q_H and (b) q_V (positive values in red and negative values in green) in units of $\frac{\theta_{oo}N^2f}{g} = 3\times10^{-7}\,\mathrm{K\,m^{-1}\,s^{-2}}$ superimposed onto the geostrophic velocity vector field at $z=0$ on day 5 (taken from Mak, 2014).

equations is virtually breaking down. As such, it signifies imminent demise of the numerical 3D fronts in this balanced model according to our interpretation. It would serve little to try to prolong the integration by means of special numerical techniques.

5.5 Energetics Analysis

It is also instructive to perform a complete analysis of the energetics of a disturbance in an SG model. Recall that the original governing prognostic equations are (2.3e,f,g).

After multiplying (2.3e) by u, (2.3f) by v and adding the resulting two equations, we would get a kinetic energy equation as

$$K_t = -(\bar{u}+\hat{u}+\tilde{u})K_x - (\hat{v}+\tilde{v})K_y - \tilde{w}K_z - \hat{u}\hat{v}\bar{u}_y - \tilde{w}\hat{u}\bar{u}_z - \hat{u}\tilde{v}\bar{u}_y + f(\hat{u}\tilde{v} - \hat{v}\tilde{u}), \quad (5.5)$$

where $K = 0.5 \times (u^2 + v^2)$ is the kinetic energy of the geostrophic flow component in the departure state per unit mass. If we multiply (2.3g) by $g\theta/\Theta_{oz}\Theta_{oo}$, we would also get an available potential energy equation. (*Note that the common notation in a discussion of energetics, P, is used here for available potential energy; this is not to be confused with the same notation for QG potential vorticity used in the 2D and 3D SG tendency equation,* (2.13) *and* (2.19).)

$$P_t = -(\bar{u}+\hat{u}+\tilde{u})\ P_x - \ (\hat{v}+\tilde{v})P_y - \tilde{w}P_z - \frac{(\hat{v}+\tilde{v})\theta g\bar{\theta}_y}{\Theta_{oo}\Theta_{oz}} - \tilde{w}\left(1 + \frac{\bar{\theta}_z}{\Theta_{oz}}\right)\frac{g\theta}{\Theta_{oo}}, \quad (5.6)$$

where $P = 0.5g \times (\theta^2/\Theta_{oo}\Theta_{oz})$ is the available potential energy of the departure state per unit mass. Furthermore, by making use of the geostrophic and hydrostatic balance relations (2.3a,b,c), we can rewrite as:

$$f(\hat{u}\tilde{v} - \hat{v}\tilde{u}) = -\nabla \cdot (\vec{V}_a\phi) + + \frac{g\tilde{w}\theta}{\Theta_{oo}}.$$

Making use of (2.3d) and the geostrophic relations, we can also rewrite the advective terms in (5.5) and (5.6) in the form of divergence of energy fluxes as follows:

$$-\tilde{u}\xi_x - \tilde{v}\xi_y - \tilde{w}\xi_z = -\nabla \cdot (\vec{V}_a\xi), \quad \vec{V}_a = (\tilde{u}, \tilde{v}, \tilde{w}), \quad \xi = K \ \ or\ P$$
$$-(\bar{u}+\hat{u})\xi_x - \hat{v}\xi_y = -\nabla \cdot (\vec{V}_g\xi), \quad \vec{V}_g = (\bar{u}+\hat{u}, \hat{v}, 0)$$

It is convenient to use more self-explanatory symbolic notation for writing the energy equations. The concise form of the energy equations are:

$$K_t = -\nabla \cdot (\vec{V}_{total}K) - \nabla \cdot (\vec{V}_a\phi) + C(\overline{K}, K) + C(P, K), \quad (5.7)$$

$$P_t = -\nabla \cdot (\vec{V}_{total}P) + C(\overline{P}, P) - C(P, K), \quad (5.8)$$

where $\vec{V}_{total} = \vec{V}_g + \vec{V}_a$, $C(\overline{K}, K) = -\hat{u}(\hat{v}+\tilde{v})\bar{u}_y - \tilde{w}\hat{u}\bar{u}_z$,

$$C(\overline{P}, P) = -(\hat{v}+\tilde{v})\theta\frac{g\bar{\theta}_y}{\Theta_{oo}\Theta_{oz}} - \tilde{w}\theta\frac{g\bar{\theta}_z}{\Theta_{oo}\Theta_{oz}} \quad \text{and} \quad C(P, K) = \tilde{w}\theta\frac{g}{\Theta_{oo}}.$$

The total energy equation is simply the sum of them:

$$E_t = -\nabla \cdot (\vec{V}_{total}E) - \nabla \cdot (\vec{V}_a\phi) + C(\overline{K}, K) + C(\overline{P}, P), \quad (5.9)$$

where $E = K + P$ is a measure of total energy per unit mass, defined in terms of the geostrophic properties of the disturbance. It should be emphasized that the total energy defined above does not include the kinetic energy of the ageostrophic velocity component, $\tilde{K}$. But $\tilde{K}$ does not vary independently, but rather must uniquely vary together with K in an SG model.

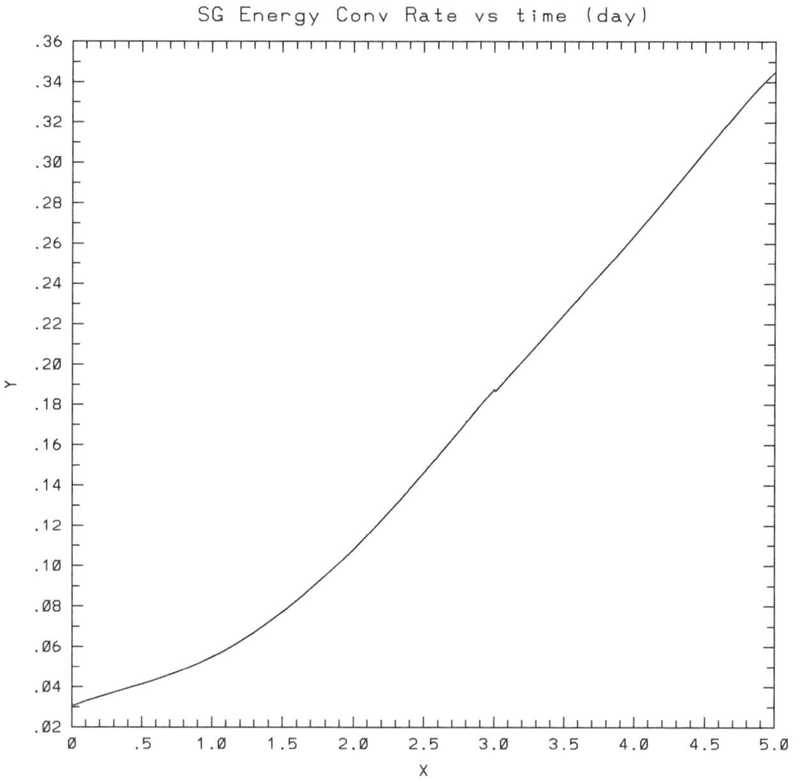

Figure 5.13. Variation of the nondimensional total energy generation rate $\langle C(\overline{K}, K) + C(\overline{P}, P)\rangle$ of the disturbance in the SG model with time in days during the evolution.

The reference state plays the role of an energy reservoir, with kinetic energy $\overline{K}$ and available potential energy $\overline{P}$ per unit mass. It is the only external energy source in support of the intensification of a disturbance. The processes of energy exchange between a disturbance of the departure state and the reference state are represented by two terms: $C(\overline{K}, K)$ and $C(\overline{P}, P)$. The former stands for the conversion rate from $\overline{K}$ to K through the Reynolds stresses acting on the shear of the reference state. The latter stands for the conversion rate from $\overline{P}$ to P arising from the eddy heat flux in the presence of the thermal gradient in the reference state. Denoting the mass integral of the fluid in a domain by an angular bracket, we would therefore have $\langle \nabla \cdot (\vec{V}_{total} E)\rangle = 0$ and $\langle \nabla \cdot (\vec{V}_a \phi)\rangle = 0$ throughout the evolution of the disturbance. The domain integrated total energy generation rate is therefore $\langle C(\overline{K}, K) + C(\overline{P}, P)\rangle$. The last row represents the rate of conversion between P and K of the disturbance. It is inconsequential as far as the total energy of the disturbance is concerned, as indicated by (5.9).

Figure 5.13 shows the variation of the domain integrated nondimensional energy conversion rate, $\langle C(\overline{K}, K) + C(\overline{P}, P)\rangle$, during the evolution. This result shows that the disturbance field intensifies monotonically with time up to day 5. Shortly after that time the frontal collapse occurs when the generalized SE equation ceases to be elliptic. The dynamic interpretation was elaborated in Section 4.4.

Appendix: Nondimensionalization of the Model Equations

It is convenient to measure the horizontal distance, vertical distance, time, and potential temperature in units of L, D, $\frac{N}{f\lambda}$ and Θ respectively. $L=\frac{ND}{f}$ is the radius of deformation. We define the nondimensional quantities as follows: $\tilde{z}=\frac{z}{D}$, $(\tilde{x},\tilde{y},\tilde{Y},\tilde{c},\tilde{b})=\frac{1}{L}(x,y,Y,c,b)$, $\tilde{t}=\frac{f\lambda}{N}t$, $\tilde{\theta}=\frac{\theta}{\Theta}$, $(\tilde{\overline{u}},\tilde{u},\tilde{v},\tilde{\hat{u}},\tilde{\hat{v}})=\frac{1}{\lambda D}(\overline{u},u,v,\hat{u},\hat{v})$, $\tilde{w}=\frac{\hat{w}N}{f_o\lambda D}$, $\tilde{\psi}=\frac{\psi N}{f_o\lambda L^2}$, $\tilde{P}=\frac{N}{f\lambda}P$, $(\tilde{\eta},\tilde{\chi})=\frac{N}{f\lambda LD}(\eta,\chi)$. A Rossby number is defined in terms of the external parameters of the reference state as $\varepsilon=\frac{\lambda}{N}=\frac{\lambda D}{fL}$. The hydrostatic balance is $\Gamma\tilde{\theta}=\tilde{\psi}_{\tilde{z}}$, with $\Gamma=\frac{g\Theta}{\lambda ND\theta_{oo}}$. The domain is $0\le\tilde{x}\le\tilde{x}_{\max}$, $0\le\tilde{y}\le\tilde{Y}$, $0\le\tilde{z}\le 1$. While the length scale of a large-scale disturbance is L, its velocity scale can be smaller than λD.

To reduce the adverse impact of the lateral boundaries, we would use a relatively narrow initial baroclinic jet. A baroclinically intensifying disturbance can be expected to be confined near the core of the jet. The following set of values of the parameters is finally adopted for the illustrative computation: $D=10^4\,\mathrm{m}$, $\lambda=4\times10^{-3}\,\mathrm{s}^{-1}$, $N=10^{-2}\,\mathrm{s}^{-1}, f=10^{-4}\,\mathrm{s}^{-1}$, $\theta_{oo}=300\,\mathrm{K}$, $\Theta=3\,\mathrm{K}$, $c=0.5(Y)$ and $b=0.8(Y)$.

The nondimensional form of (2.22) and (2.23) can then be written as:

$$\begin{aligned}
&S_1\{\eta\}+R_1\{\chi\}=G\{\psi\}\\
&S_1 \quad =(\varepsilon^{-1}+\hat{v}_z)\frac{\partial^2}{\partial z^2}-2\hat{v}_z\frac{\partial^2}{\partial x\partial z}+(\varepsilon^{-1}+\psi_{zz})\frac{\partial^2}{\partial x^2}\\
&R_1 \quad =\hat{v}_y\frac{\partial^2}{\partial z^2}+(\varepsilon^{-1}+\psi_{zz})\frac{\partial^2}{\partial x\partial y}-\hat{v}_z\frac{\partial^2}{\partial y\partial z}+(\overline{u}_z+\hat{u}_z)\frac{\partial^2}{\partial x\partial z}\\
&G\{\psi\} \quad =2(\overline{u}_z+\hat{u}_z)\hat{v}_x+2\hat{v}_y\hat{v}_z,
\end{aligned}\tag{5.A1}$$

$$\begin{aligned}
&S_2\{\chi\}+R_2\{\eta\}=H\{\psi\}\\
&S_2 \quad =(\varepsilon^{-1}-\overline{u}_y-\hat{u}_y)\frac{\partial^2}{\partial z^2}+2(\overline{u}_z+\hat{u}_z)\frac{\partial^2}{\partial y\partial z}+(\varepsilon^{-1}+\psi_{zz})\frac{\partial^2}{\partial y^2}\\
&R_2 \quad =-\hat{u}_x\frac{\partial^2}{\partial z^2}+(\varepsilon^{-1}+\psi_{zz})\frac{\partial^2}{\partial x\partial y}-\hat{v}_z\frac{\partial^2}{\partial y\partial z}+(\overline{u}_z+\hat{u}_z)\frac{\partial^2}{\partial x\partial z}\\
&H\{\psi\} \quad =-2(\overline{u}_z+\hat{u}_z)\hat{u}_x-2(\overline{u}_y+\hat{u}_y)\hat{v}_z.
\end{aligned}\tag{5.A2}$$

The boundary conditions for η are $\eta=0$ at $y=0,Y$ and $z=0,1$. With such nondimensional form of the operators, (2.15) and (2.16) may also be viewed as expressions for the nondimensional χ and η.

The nondimensional form of (2.4) for the reference state in units of $\frac{\theta_{oo}N^2f}{g}$ is:

$$\overline{q}=1-\varepsilon\overline{u}_y-\varepsilon^2(\overline{u}_z)^2. \tag{5.A3}$$

The SG PV consists of two parts ($q = q_H + q_V$ where q_H and q_V are associated with the horizontal and vertical components of the absolute vorticity respectively). Their nondimensional forms are power series of the Rossby number ε, namely:

$$q_H = -\varepsilon^2((v^*_z)^2 + (u^*_z)^2) + \varepsilon^3(v^*_z(u^*_y v^*_z - u^*_z v^*_y) - u^*_z(u^*_z v^*_x - u^*_x v^*_z)) \tag{5.A4}$$

$$q_V = 1 + \varepsilon(v^*_x - u^*_y + \psi^*_{zz}) + \varepsilon^2(u^*_x v^*_y - u^*_y v^*_x + \psi^*_{zz}(v^*_x - u^*_y)) + \varepsilon^3(\psi^*_{zz}(u^*_x v^*_y - u^*_y v^*_x)). \tag{5.A5}$$

To diagnose the impending frontal collapse toward the end of integration, we will examine q_H and q_V separately.

Recap

- It is more straightforward to analyze 3D SG frontal dynamics in physical coordinates than in geostrophic coordinates.
- Geostrophic streamfunction is governed by a single prognostic 3D SG tendency equation.
- The Sawyer–Eliassen equation is generalized to two coupled diagnostic equations that relate two ageostrophic potentials to the geostrophic streamfunction at each time instant.
- Formal analytic solutions for the ageostrophic potentials can be written as two separate functions of the geostrophic streamfunction.
- An illustrative analysis has been performed using a baroclinic jet as the reference state. The geostrophic streamfunction of the most unstable 3D Eady mode and a corresponding ageostrophic circulation is used as an initial disturbance.
- Realistic cold fronts and warm fronts are simulated at the same time in a few days.
- Dynamic difference between the cold front and warm front stems from the fact that the roles of the geostrophic and ageostrophic velocity components are not completely separable. They jointly contribute to the cross-front scale contraction as well as to the intensification of the thermal contrast. Both components give rise to cold advection in the cold front region. In contrast, the geostrophic (ageostrophic) velocity component gives rise to warm (cold) advection in the warm front region.
- Mature fronts emerge in a few days; strong shear and weak stratification in the frontal region would lead to a breakdown of the generalized Sawyer–Eliassen equations. They eventually become no longer elliptic partial differential equations, leading to frontal collapse in the SG model framework.

6

Impact of Latent Heat Release on SG Frontal Dynamics

6.1 Introductory Remarks

There is typically considerable precipitation in the cold and warm frontal regions of an intensifying mid-latitude cyclone. The related cloud field and meso-scale convective systems would release enormous amounts of latent heat. One would expect a substantial impact of such condensational heating on the development of the fronts and cyclone as a whole. A positive impact would mean that there exists a symbiotic relationship among the innumerable cloud elements and the cyclone–front system. In such a relationship, the large-scale wind field in the frontal regions would bring about the moisture supply for the convection and the self-induced latent heat release would in turn strengthen the gradient of the pressure field and thereby the wind field itself. From an energetic point of view, the intensification would be enhanced if the heating on average takes place in the warmer sector of a cyclone. More gravitational potential energy would then be generated and part of that in turn would be converted to kinetic energy of the wind field. Evidence of such a positive correlation is commonly observed. A mathematical treatment of the related processes could be very complex, depending upon how much detail one is prepared to deal with.

The objective of this chapter is to present two illustrative analyses that quantify the positive feedback effect of latent heat release upon the evolution of fronts. We first examine the problem in a 2D semi-geostrophic (SG) setting and then in a 3D SG setting. The results of each moist SG model and the corresponding dry model will be compared. In order to make the discussions in this chapter self-contained, we will outline substantial details of each of the two mathematical analyses. Only if the readers wish to know about the detailed background of the dry model simulations would it be necessary to read Chapters 2–5. We will use a general model incorporating the diabatic heating processes to get more reliable quantitative results for the impacts in Chapter 7.

6.2 Condensational Heating Parameterization

Condensational heating in the atmosphere stems from the phase changes of the water substance in raindrops and snow/ice crystals subject to the influence of the turbulent motions in a cloud field. These complicated processes are messy in details and are daunting to deal with. In order to translate conceptual notions about such subgrid-scale processes to a manageable mathematical representation of condensational heating, it is necessary to introduce broad idealizations. If the model grid resolution is sufficiently fine and the non-hydrostatic effect is retained, one would not need to invoke a convection parameterization scheme at all and would only require a scheme for the microphysics of precipitation in order to compute the latent heat release. One such application will be discussed in Chapter 7. However, for the purpose of delineating the basic aspect of moist dynamics in the context of a basic model, it would suffice to apply a convection parameterization scheme.

There exist a fair number of convection parameterization schemes, which differ in their degrees of complexity. However complicated a scheme might appear to be, it would still be at best a coarse proxy of what actually happens in a precipitating atmosphere. The degree of simplicity in a chosen parameterization scheme should be commensurate with the level of approximation in a dynamical model framework. For the purpose of deducing a theoretical understanding of the impact of latent heat release on frontal dynamics, it would be preferable to use the simplest possible rather than an existing more comprehensive parameterization scheme. In a simple scheme for latent heat release, the water substance budget is not explicitly represented in the model. Rather, one would seek to express the heating rate as a function of one large-scale field variable with a few supplemental parameters in an equivalent dry model.

Condensational heating is represented by using a smaller (larger) value of stratification where the vertical motion is upward (downward) in one simple scheme. This is based on the assumption that updraft is saturated, whereas downdraft has no water in it. The mathematical form of this parameterization is based on the idea of conservation of dry/moist entropy under a pseudo-adiabatic assumption, known as the EFT scheme (Emanuel et al., 1987). The thermodynamic equation would be then $\frac{d\ln\theta}{dt} = 0$ for areas of descent $\frac{d\ln\theta}{dt} = w\left[\frac{\partial\ln\theta}{\partial Z} - \frac{\Gamma_m}{\Gamma_d}\frac{\partial\ln\theta_e}{\partial Z}\right]$ for areas of ascent where Z is geostrophic height coordinate, $\frac{d}{dt} = \frac{\partial}{\partial t} + \vec{V}\cdot\nabla$, $\vec{V}$ is the velocity, Γ_m is the moist adiabatic lapse rate and Γ_d is the dry adiabatic lapse rate. The other notations have the usual meanings. Those investigators applied this scheme in an analysis of moist 2D SG frontogenesis starting with an unstable Eady modal disturbance in geostrophic coordinates. It has also been applied in a study of transient frontogenesis with a 2D SG model in geostrophic coordinates starting with a nonmodal disturbance that has nonzero interior potential vorticity (Montgomery and Farrell, 1991).

Lapeyre and Held (2004) devised a moist two-layer quasi-geostrophic (QG) beta-plane model in which moisture content is represented by a proxy variable. An essentially moist-adjustment scheme is used to parameterize the latent heat release. They used such a model to investigate moist baroclinic eddy dynamics. Jets and waves dominate the model flow when the model heating is weak, somewhat similar to dry eddy dynamics. Their calculation also brings to light a vortex-dominated flow regime when the intensity of heating is sufficiently increased. The treatment was interpreted in terms of an effective static stability. It corresponds to one deduced with weighting between saturated and unsaturated regions in a moist environment. O'Gorman (2011) devised a variation of this version of *effective static stability* scheme. He used an empirically deduced latitude-height dependent rescaling coefficient in conjunction with the eddy vertical velocity in a multi-level primitive equation global model. The statistical structure of the atmospheric general circulation is reasonably well simulated with such a parameterization scheme without an explicit hydrologic cycle. Such a scheme has not been applied in a simulation study of frontogenesis.

In another simple scheme, the source of moisture for convection is assumed to be a preexisting surface moisture layer not unlike the atmosphere. It follows that there would be frictionally induced moisture convergence. An upward flux of moisture into the interior is assumed to take place proportional to the cyclonic vorticity at the top of a frictional layer in accordance with Ekman layer dynamics. It would be necessary to specify a supplemental vertical distribution of the heating in this treatment. This is the basic notion of Conditional Instability of the Second Kind (CISK). This scheme was originally devised in a linearized form for an instability analysis of hurricane-like disturbance (Charney and Eliassen, 1964). This parameterization scheme has also been applied to analytic studies of moist baroclinic instability (e.g., Mak, 1994) and of 2D SG frontogenesis associated with a surface disturbance driven by a prescribed deformation-wind field performed in geostrophic coordinates (Mak and Bannon, 1984). It should be stressed that the concept of CISK is physically meaningful no matter whether the moisture comes from evaporation over a local water surface or from a preexisting large-scale environmental moist boundary layer. The relative importance of the two sources to the development of a tropical cyclone obviously depends on the particular circumstance.

An important feature of condensational heating is that it is positive-definite. It means that each spectral component in the flow field of a disturbance would induce a contribution to a heating field consisting of not only that spectral component itself, but also all of its higher harmonics. In other words, the impact upon the flow field by such a self-induced condensational heating would be intrinsically nonlinear. This feature of heating should be present in whichever parameterization scheme one chooses to apply.

The moist frontal model analyses in this chapter are performed in *physical coordinates* rather than in geostrophic coordinates. The advantage of working with physical coordinates was discussed in Section 2.4.1. The following mathematical formula represents the parameterized condensational heating rate:

$$Q=\alpha h D\{\zeta^*\}$$

$$D\{\zeta^*\}=\begin{cases}\zeta^*=\nabla^2\psi^* & if \quad \zeta^*>0\\ 0 & if \quad \zeta^*<0\end{cases}, \tag{6.1}$$

where ψ is the geostrophic streamfunction, $\nabla^2 = \frac{\partial^2}{\partial x^2} + \frac{\partial^2}{\partial y^2}$. It is in units of K s^{-1}. The asterisk * in (6.1) refers to a height corresponding to the top level of an implicit moist boundary layer in an equivalent dry model. The horizontal distribution of the heating is controlled by that of $D\{\zeta^*\}$ and the vertical distribution of the heating by a prescribed heating profile $h(z)$, $0 \le h(z) \le 1$. The strength of the heating is introduced through a single parameter α in unit K. This simple parameterization is very crude indeed, but it does rest upon sound dynamic consideration.

6.3 Two-Dimensional SG Frontal Model with Condensational Heating

We begin by experimenting with the heating parameterization (6.1) in the context of the simplest possible 2D SG model. The nature of dry frontal dynamics in this setting has been thoroughly examined in the illustrative analysis presented in Section 4.2. In other words, frontogenesis in this model analysis is associated with additional potential vorticity anomalies arising from the impact of release of latent heat. This experiment would help develop a feel for the different elemental aspects of the impact of condensation heating on frontal dynamics. The results of this moist model analysis will be compared with the counterpart results reported by Emanuel et al. (1987). Recall that the total state in a model at any time is the sum of a reference state and a time-dependent departure state. The reference state under consideration is characterized by a stable stratification $N = \sqrt{\frac{g}{\theta_{oo}} \frac{d\theta_o}{dz}} > 0$ and a zonal baroclinic shear flow $\overline{u}(z)$, which is in thermal wind balance with an additional component in the potential temperature field $\overline{\theta}(y,z)$.

We use the same notations as in Chapter 2, in that the geostrophic wind and the related potential temperature are indicated by a superscript ^ and the ageostrophic wind component $(\tilde{u}, \tilde{w})$ by a superscript ~ . The governing equations in this case are then:

$$\frac{D\hat{v}}{Dt} = -f\tilde{u}, \tag{6.2a}$$

$$\frac{D(\overline{\theta} + \hat{\theta})}{Dt} + \hat{v}\overline{\theta}_y + \tilde{w}\frac{d\theta_o}{dz} = \alpha h D\{\hat{v}_x^*\}, \tag{6.2b}$$

$$\tilde{u}_x + \tilde{w}_z = 0, \tag{6.2c}$$

$$\hat{\theta}_x = \frac{\theta_{oo} f}{g} \hat{v}_z, \tag{6.2d}$$

where $\frac{D}{Dt} = \frac{\partial}{\partial t} + (\overline{u} + \tilde{u})\frac{\partial}{\partial x} + \tilde{w}\frac{\partial}{\partial z}$ is the total derivative. The domain is $-\infty < x < \infty$, $0 \le z \le D$. The boundaries $z = 0, D$ are rigid surfaces through which there is no mass flux. We set $z = 0.1$ as the top of an implicit moist layer indicated by the asterisk in (6.2b).

As before, we introduce a geostrophic streamfunction ψ so that $\hat{v}=\psi_x$ and $\hat{\theta}=\frac{f\theta_{oo}}{g}\psi_z$. Its local rate of change can be evaluated with the use of a *2D SG tendency equation* derived in Section 2.3.3 with an additional term that represents the influence of self-induced heating, namely:

$$P_t=-(\bar{u}+\tilde{u})P_x-\tilde{w}P_z-\tilde{u}_x\psi_{xx}-\tilde{w}_z\frac{f^2}{N^2}\psi_{zz}-\left(\tilde{w}_x+\tilde{u}_z\frac{f^2}{N^2}\right)\psi_{xz}+\frac{fg\alpha}{N^2\theta_{oo}}h_zD\{\overset{*}{\psi}_{xx}\}. \tag{6.3}$$

The condition $\hat{w}=0$ at $z=0,D$ reduces (6.2b) to

$$@z=0, \quad D\psi_{zt}=-(\bar{u}+\tilde{u})\psi_{xz}-\psi_x\bar{\theta}_y. \tag{6.4}$$

P is identifiable with the quasi-geostrophic potential vorticity. Equation (6.3) differs from the counterpart equation of a dry model (2.13) in having the additional term on the RHS that represents the effect of heating. We also introduce an ageostrophic potential χ such that $\tilde{u}=-\chi_z$ and $\tilde{w}=\chi_x$. The Sawyer–Eliassen (SE) equation relates the ageostrophic potential to the geostrophic streamfunction under the influence of heating as:

$$f(f+v_x)\chi_{zz}-\frac{2g\theta_x}{\theta_{oo}}\chi_{xz}+\left(N^2+\frac{g\theta_z}{\theta_{oo}}\right)\chi_{xx}=2f\bar{u}_zv_x+\frac{g\alpha}{\theta_{oo}}hD\{\overset{*}{\psi}_{xxx}\}. \tag{6.5}$$

The appropriate boundary conditions are $\chi(x,0,t)=\chi(x,D,t)=0$.

We nondimensionalize the equations, as in the dry model analysis detailed in Section 2.6. The nondimensional heating parameter is $\tilde{\alpha}=\frac{g\alpha}{N\lambda D\theta_{oo}}$. The nondimensional set of the governing equations used in doing the computation is then (without the tilde notation):

$$P_t=-(\bar{u}+\tilde{u})P_x-\tilde{w}P_z-\tilde{u}_x\psi_{xx}-\tilde{w}_z\psi_{zz}-(\tilde{w}_x+\tilde{u}_z)\psi_{xz}+\alpha h_zD\{\overset{*}{\psi}_{xx}\}, \tag{6.6}$$

$$P=\psi_{xx}+\psi_{zz}, \tag{6.7}$$

with

$$@z=0, \quad D\psi_{zt}=-(\bar{u}+\tilde{u})\psi_{xz}+\psi_x\bar{u}_z. \tag{6.8}$$

The nondimensional SE equation with heating would be

$$(\varepsilon^{-1}+v_x)\chi_{zz}-2\psi_{xz}\chi_{xz}+(\varepsilon^{-1}+\psi_{zz})\chi_{xx}=2\bar{u}_zv_x+\alpha hD\{\overset{*}{\psi}_{xxx}\}\}, \tag{6.9}$$

with $\chi(x,0,t)=\chi(x,D,t)=0$.

6.3.1 Results with a Generic Heating Profile

The variables of this model are nondimensionalized in the same way as in the dry SG model elaborated in Section 5.6. There is now one additional nondimensional parameter $\tilde{\alpha}=\frac{g\alpha}{N\lambda D\theta_{oo}}$ for prescribing the strength of the model condensational heating.

There would be one extra term on the RHS of (5.A1) associated with the heating, namely $\alpha h_z D\{\varsigma^*\}$. Equations (5.A2) and (5.A3) would not change, except that $z = 0$ is set to the level indicated by the asterisk in (6.18). Likewise, there would also be an extra term in $G\{\psi\}$ associated with the heating in (5.A4), namely $\gamma\alpha h \frac{\partial}{\partial x} D\{\varsigma^*\}$, and in $H\{\psi\}$ in (5.A5), namely $-\gamma\alpha h \frac{\partial}{\partial y} D\{\psi\}$.

The algorithm used in Section 4.2 is applied here to integrate the system of equations, from (6.6) to (6.9), for the case of a reference state that has an Eady flow, $\overline{u} = \lambda z$. The corresponding potential temperature of the reference state is $(\theta_o + \overline{\theta})$, where $\overline{\theta} = -\frac{f\theta_{oo}\lambda}{g} y$ and $(\theta_o)_z = \frac{\theta_{oo}}{g} N^2$. First, let us experiment with a generic heating profile, which is symmetric about the mid-level of the model atmosphere:

$$h(z) = \cos(0.5*\pi*((z - z_1)/\Delta)), \tag{6.10a}$$

where $z_1 = 0.5, \Delta = 0.5$. Furthermore, we use $\alpha = 1$ and the same values for all other parameters as in the dry case.

A self-consistent initial disturbance is the most unstable Eady mode shown in Figure 4.1. The integration reveals that the self-induced heating strongly enhances the frontogenetic process. A systematic comparison of the structure of four field variables of the flow with versus without heating would serve to highlight the nature of the moist frontal dynamics. First, the potential temperature of the disturbance with versus without heating on day 2.5 is compared in Figure 6.1a,b. Under the influence of self-induced heating, the fronts at both bottom and top surfaces are considerably more intense and the frontal surface has a greater slope. The surface front is also located further east, implying that the developing cyclone–front system propagates faster eastward under the influence of heating. In contrast to the case of the dry model, the θ field is no longer symmetric about the mid-level even though the heating profile itself is symmetric. The reason is that $\frac{dh}{dz}$ is asymmetric about the mid-level. The heating affects the potential vorticity in the lower half of the domain differently than in the upper half of the domain. The impact is greater where we have $\frac{dh}{dz} > 0$ than where we have $\frac{dh}{dz} < 0$. At the levels in the lower half of the domain, where the heating increases with height, the process of vortex stretching would be enhanced. Consequently, such distribution of heating would contribute to an increase in the rate of change of the potential vorticity field in accordance with (6.6). The enhanced low-level vorticity would in turn induce stronger heating according to this parameterization. This positive feedback underlies the symbiotic interaction between the flow field and the heating, as manifested in Figure 6.1b. The bottom surface frontogenesis is much more enhanced than the top surface frontogenesis. Figure 6.1b is quite similar to the counterpart result that Emanuel et al. (1987) obtained with the EFT scheme (see panel (a) of their figure 7). The frontal collapse occurs about eight hours sooner in the moist model.

The corresponding geostrophic velocity fields with versus without heating are compared in Figure 6.2a,b. The difference is equally pronounced. Although the gradient of this along-front velocity is already quite pronounced in the dry model, it is

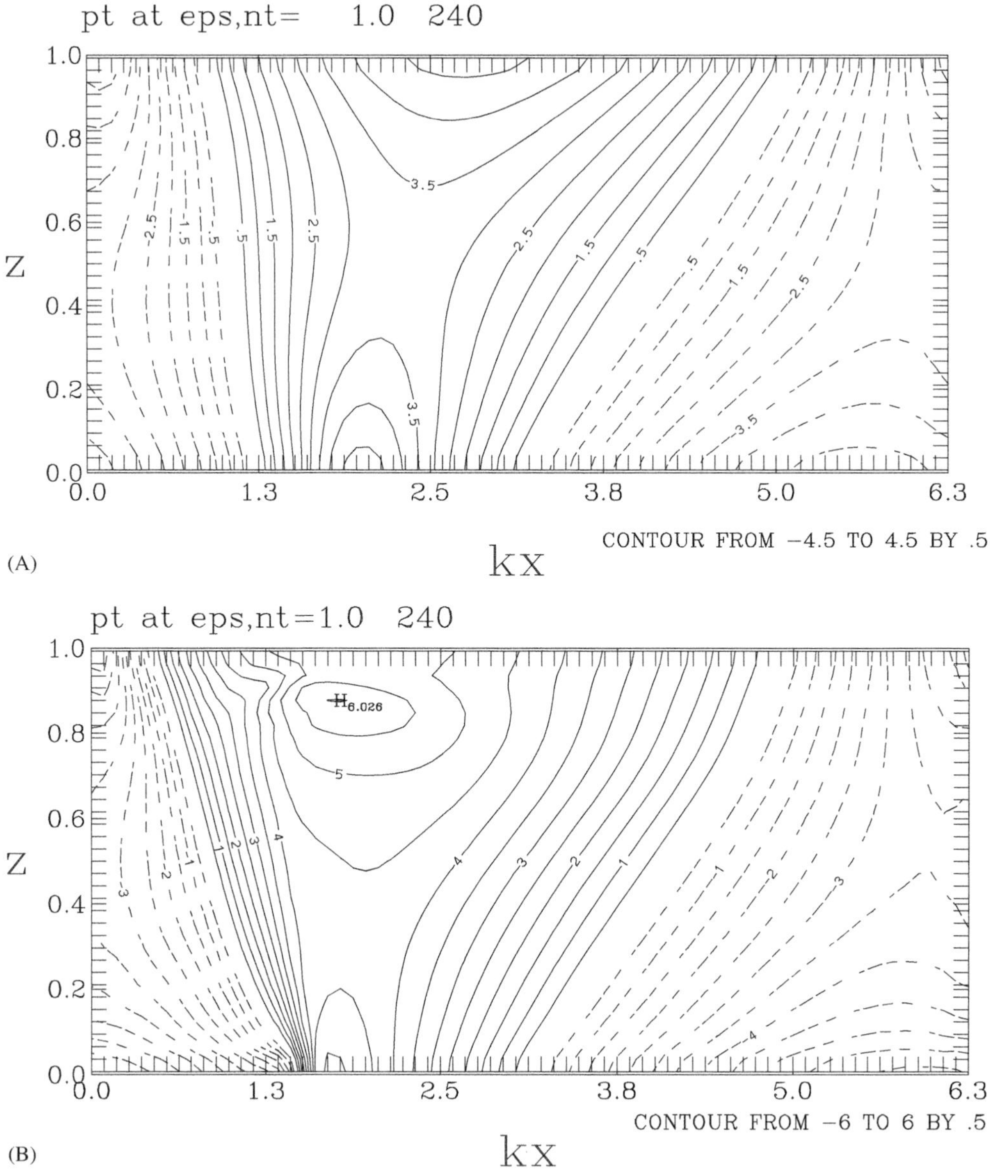

Figure 6.1. Structure of the nondimensional disturbance potential temperature field (a) without heating and (b) with heating on day 2.5 in the case of an Eady flow in the reference state.

even more so in this moist model. Note that the contour interval in panel (b) is 2.5 times larger than in panel (a). The two plots would look more dramatically different had the same value of CI been used. The vertical tilt of v in the lower half of the model domain is distinctly smaller than that next to the upper boundary surface. The stronger cyclonic vorticity at the surface signifies a mutual reinforcement between the heating and the intensification of the disturbance.

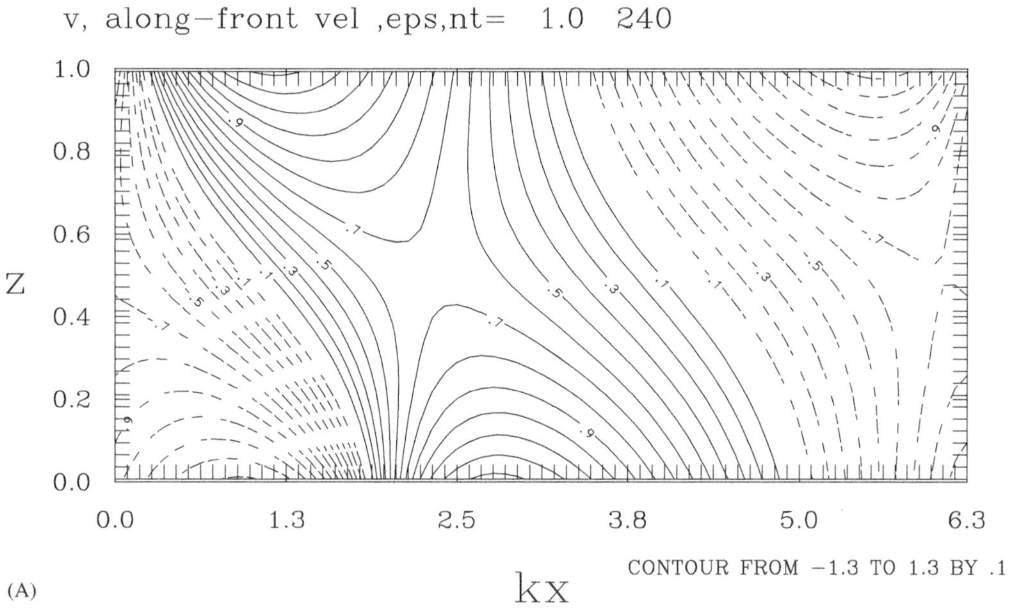

(A)

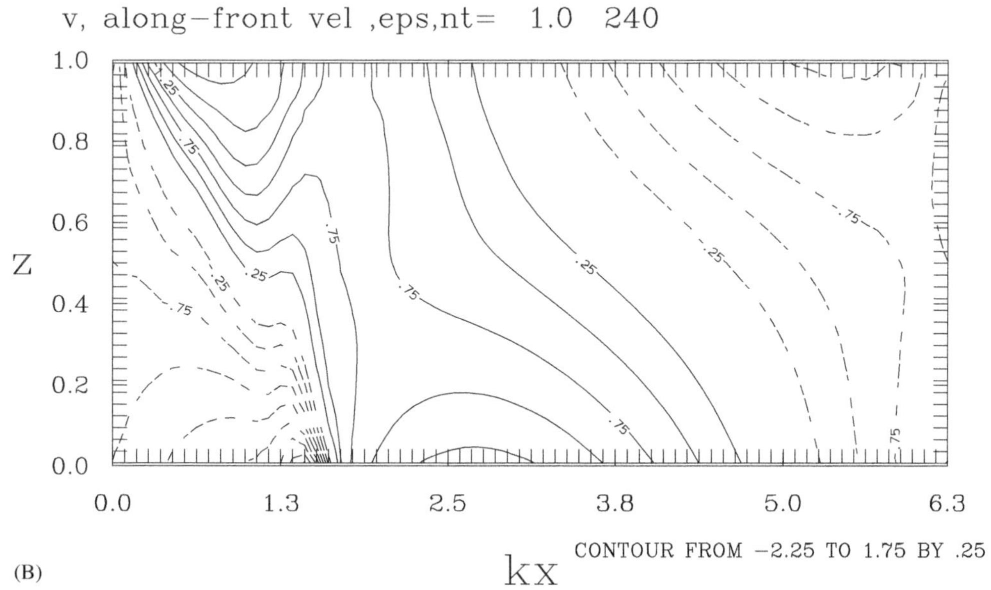

(B)

Figure 6.2. Structure of the nondimensional geostrophic velocity field (a) without heating and (b) with heating on day 2.5 in the case of an Eady flow in the reference state.

A corresponding comparison of the ageostrophic potential fields with versus without heating on day 2.5 is presented in Figure 6.3a,b. In the absence of heating, the intensity of the two cells is equally strong for the range of values of χ from –0.065 to 0.065, but the cell with negative values of χ is considerably narrower than the one with positive values. It follows that the maximum descent is equally strong

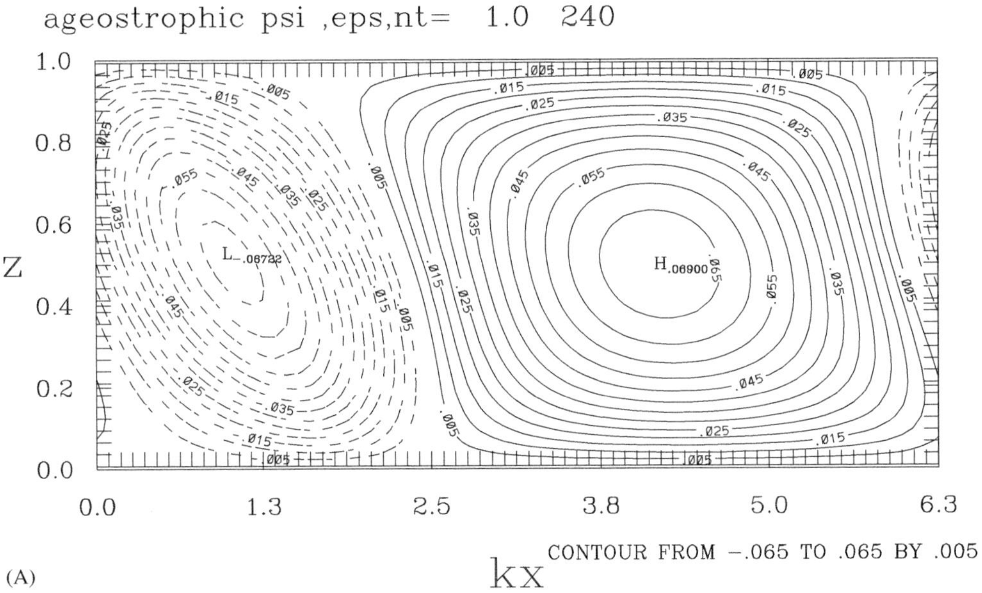

(A)

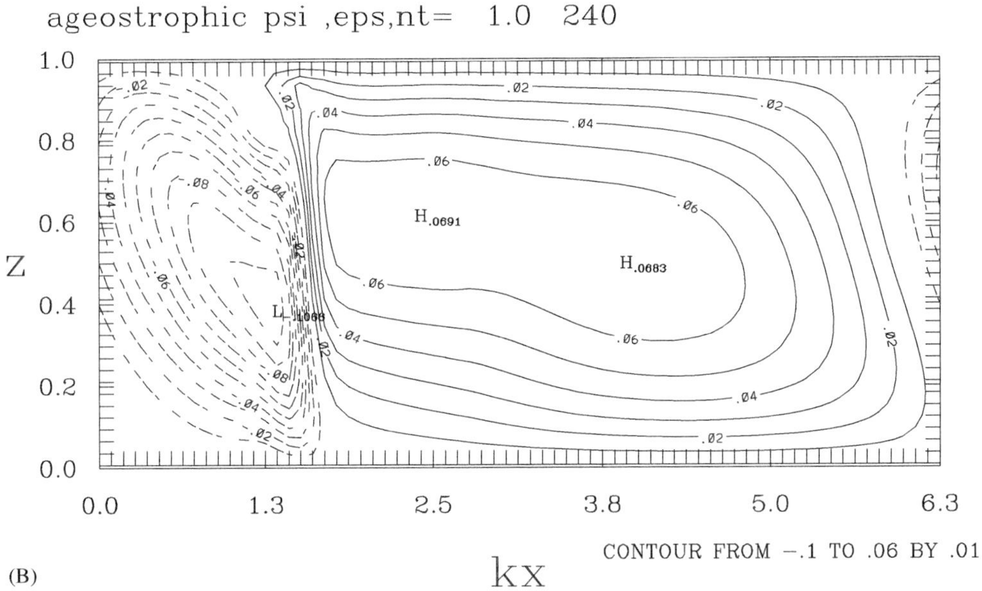

(B)

Figure 6.3. Structure of the nondimensional ageostrophic potential field (a) without heating and (b) with heating on day 2.5 in the case of an Eady flow in the reference state.

to the maximum ascent because the area of ascent is the same as the area of descent (panel a). With heating, however, the two cells are changed differently, not only in their shape but also in their intensity. The values of χ now range from –0.1 to 0.06. Note again that the CI in panel (b) is twice as large as that in panel (a). The shapes of the two cells are greatly modified in the region of ascent where latent heat release

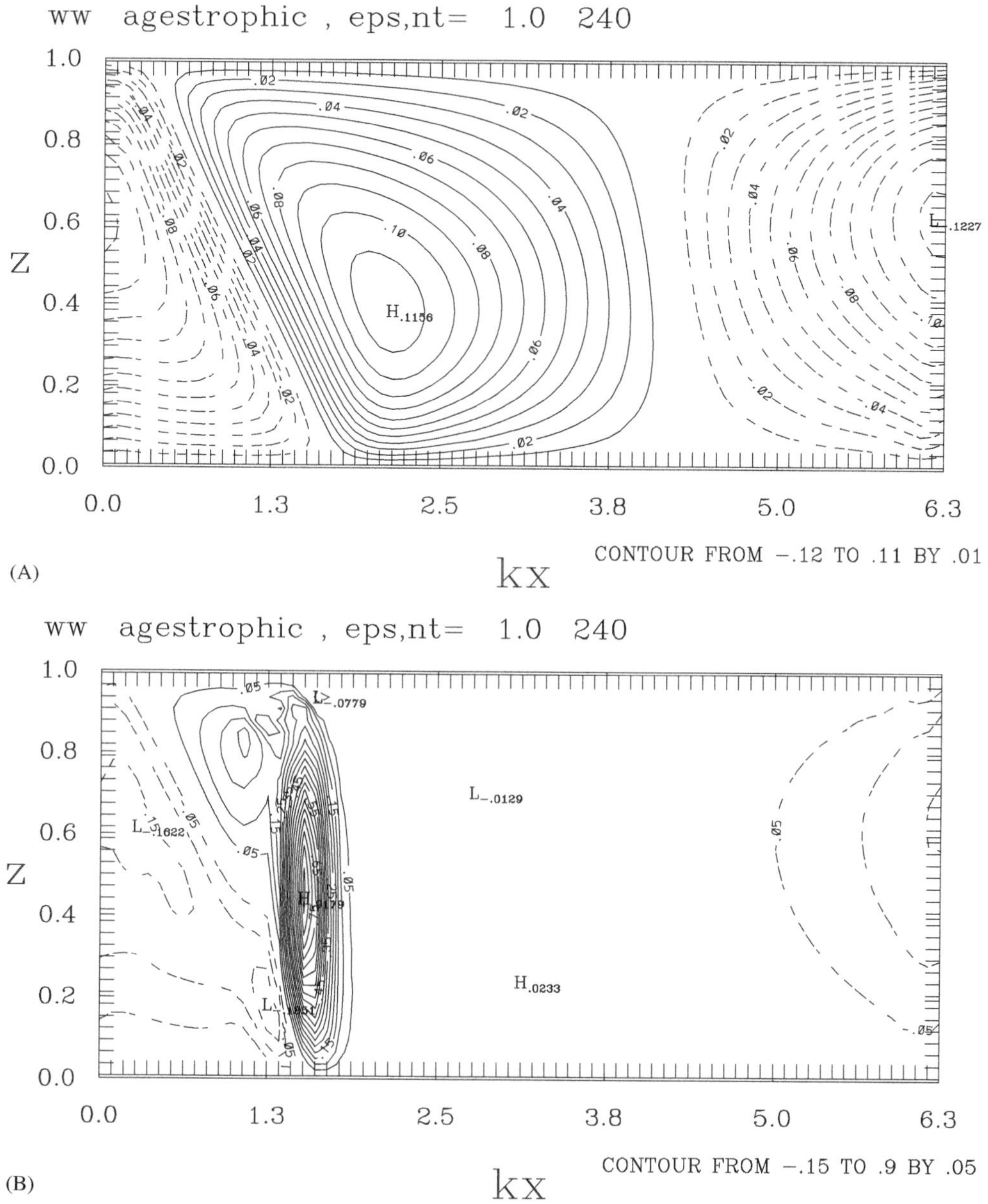

Figure 6.4. Structure of the nondimensional vertical velocity field (a) without heating and (b) with heating on day 2.5 in the case of an Eady flow as the reference state.

takes place. This result is similar to the counterpart result of Emanuel et al. (1987; see panel (c) of their figure 7).

The impact of heating is most evident in the vertical velocity field, as seen in Figure 6.4a,b. The ascent is concentrated over the surface front and has virtually no tilt. The maximum ascent is nearly eight times stronger under the influence of heating, located over the surface front. The area of ascent is much smaller than the area of descent. It is so narrow that the contours are tightly packed together even though the CI in panel (b) is five times larger than that in panel (a). It follows that the descent is much weaker and spread over a much larger area. This result is similar to the counterpart result of Emanuel et al. (1987; see panel (d) of their figure 7).

In light of such highly accentuated distribution of all frontal properties in the flow field, it is no wonder that frontal collapse occurs sooner with heating, shortly after day 2.5. At that time, the model resolution is simply no longer capable of resolving the structure for further development of the disturbance. Computational noise soon overwhelms the fields and the integration stops.

6.3.2 Dependence on Heating Profile

The vertical distribution of condensational heating is an internal property of a precipitating cloud field in the atmosphere. It is a property very difficult to deduce observationally, even from ground- or satellite-based radar data. There is unlikely to be a universal heating profile. A particular shape would depend on the individual circumstances, such as the vertical distributions of the large-scale moisture, temperature, and wind, as well as background stratification. It is therefore not warranted to put undue emphasis on the merit of any particular heating profile. It would be more meaningful to primarily quantify the sensitivity of the impact to a range of plausible heating profiles. So let us repeat the integration using two asymmetric heating profiles that have substantially different levels of maximum value, namely.

$$h(z) = \exp\left\{-\left(\frac{z-z_*}{\Delta}\right)^2\right\}, \tag{6.11}$$

with $\Delta = \begin{cases} \Delta_1 \, for \; z > z_* \\ \Delta_2 \, for \; z < z_* \end{cases}.$

The level of maximum heating is z_*. It suffices to ascertain the differences in the impacts for two sample members of this class:

(a) Profile LOW: $z_* = 0.25, \Delta_1 = 0.5, \Delta_2 = 0.15$
(b) Profile HIGH: $z_* = 0.75, \Delta_1 = 0.15, = 0.5$.

Figure 6.5 shows these two heating profiles, one having larger values in the lower troposphere and the other in the upper troposphere.

The structure of four properties of the disturbance θ, v, χ, w on day 2.5 associated with these two heating profiles is compared in Figures 6.6–6.9. The differences are substantial. But by and large, the two responses are not drastically dissimilar, confirming that as far as this model is concerned, the sensitivity of frontal dynamics to the heating profile is not too great. The results establish that the surface front under the influence of heating profile LOW is more intense than under the influence of heating profile HIGH. This is particularly true in the ageostrophic circulation (Figures 6.8 and 6.9). This is to be expected, in light of our previous discussion of

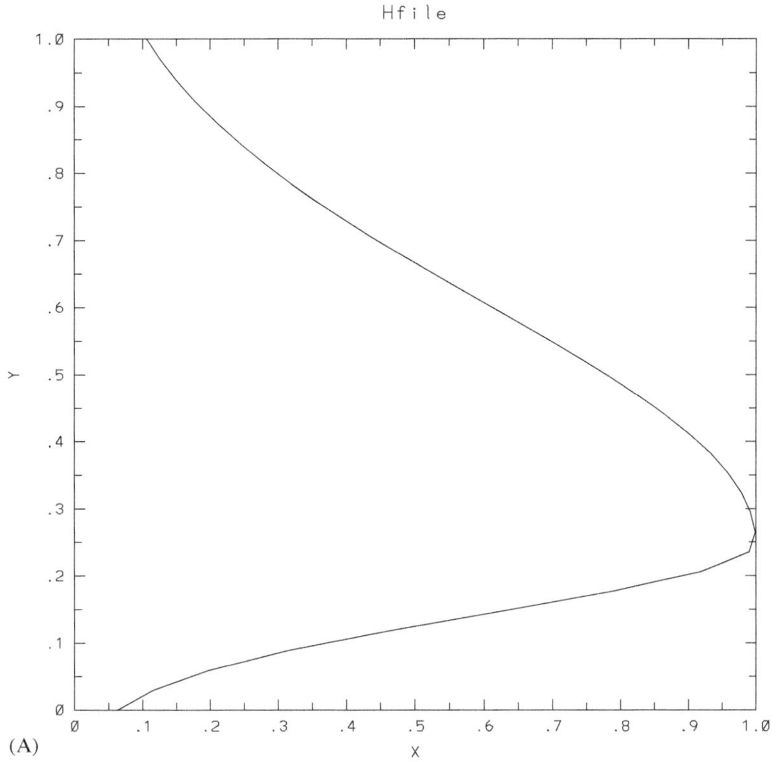

(A)

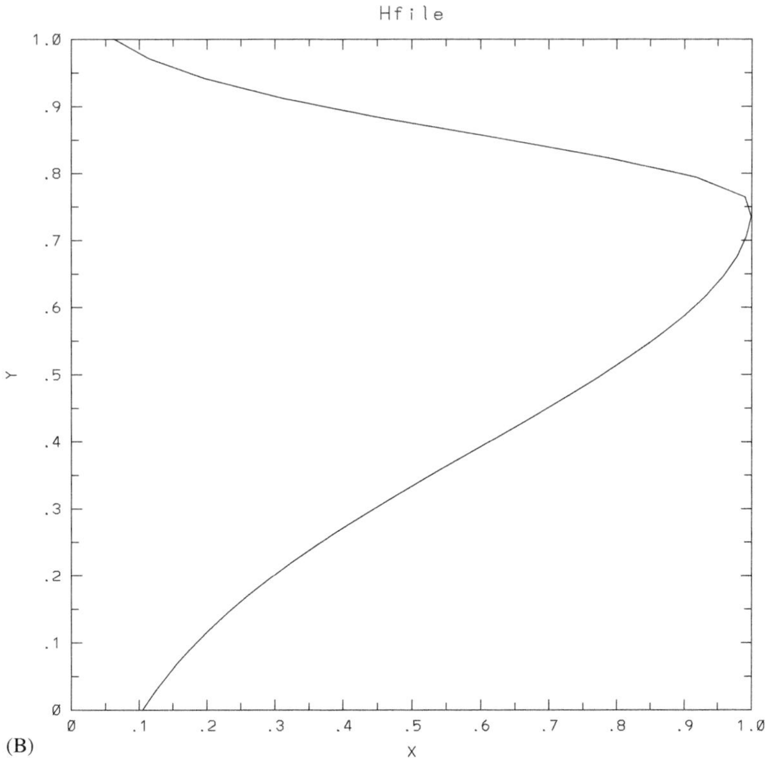

(B)

Figure 6.5. Heating profiles $h(z)$ referred to as (a) profile LOW and (b) profile HIGH.

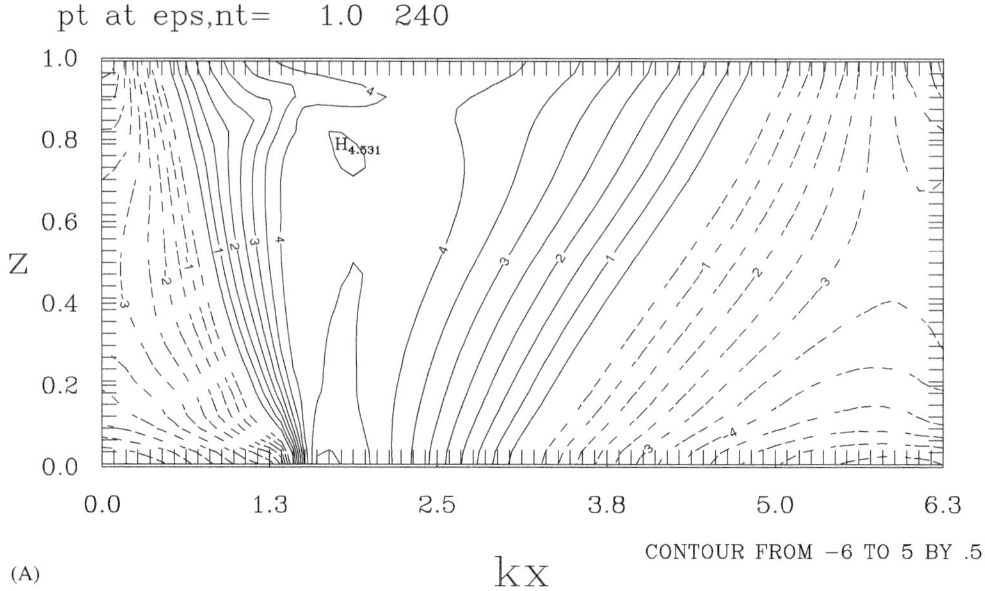

(A)

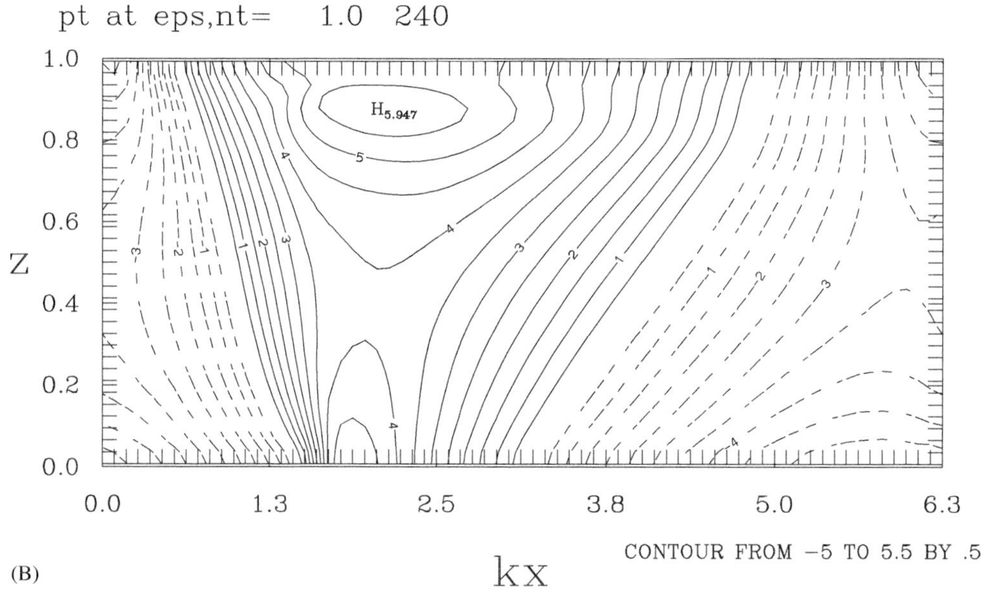

(B)

Figure 6.6. Structure of the nondimensional potential temperature field on day 2.5 for (a) profile LOW and (b) profile HIGH in the case of an Eady flow as the reference state.

Figure 6.3, because $\frac{dh}{dz}$ at the lower tropospheric levels has larger positive values in the case of profile LOW. It follows that the positive feedback effect must be stronger. The main point is that the impact on the frontal formation and intensity by a self-induced condensational heating is quite substantial for all three prototype heating profiles.

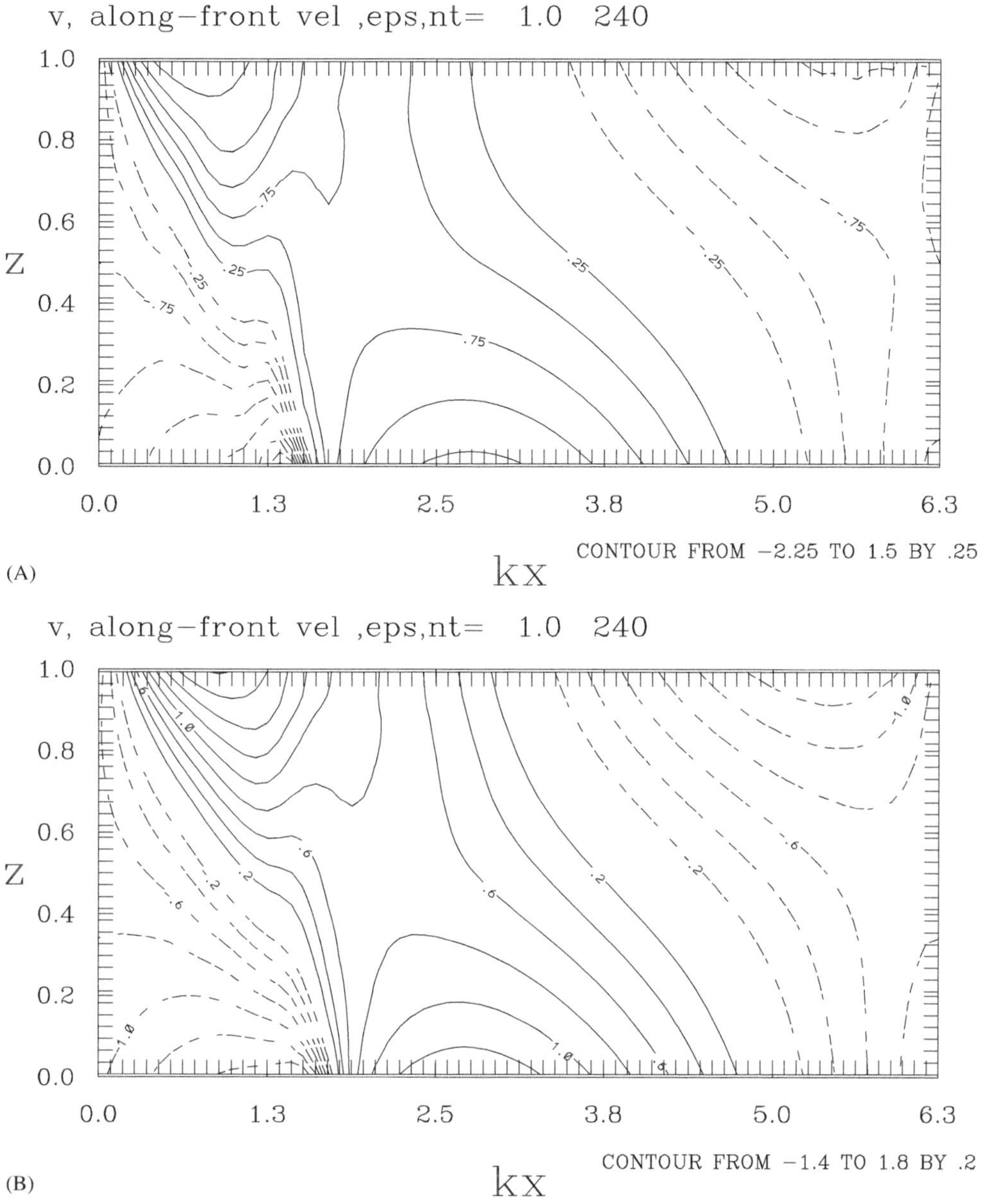

Figure 6.7. Structure of the nondimensional geostrophic velocity field on day 2.5 for (a) profile LOW and (b) profile HIGH in the case of an Eady flow as the reference state.

6.4 Three-Dimensional SG Frontal Model with Condensational Heating

Next we perform an illustrative analysis with a 3D SG model so that we might ascertain the impact of latent heat release on frontal dynamics with identical parameterized heating in a dynamically less restrictive setting. The results of the illustrative analysis for the dry 3D SG model presented in Chapter 5 will be used as a

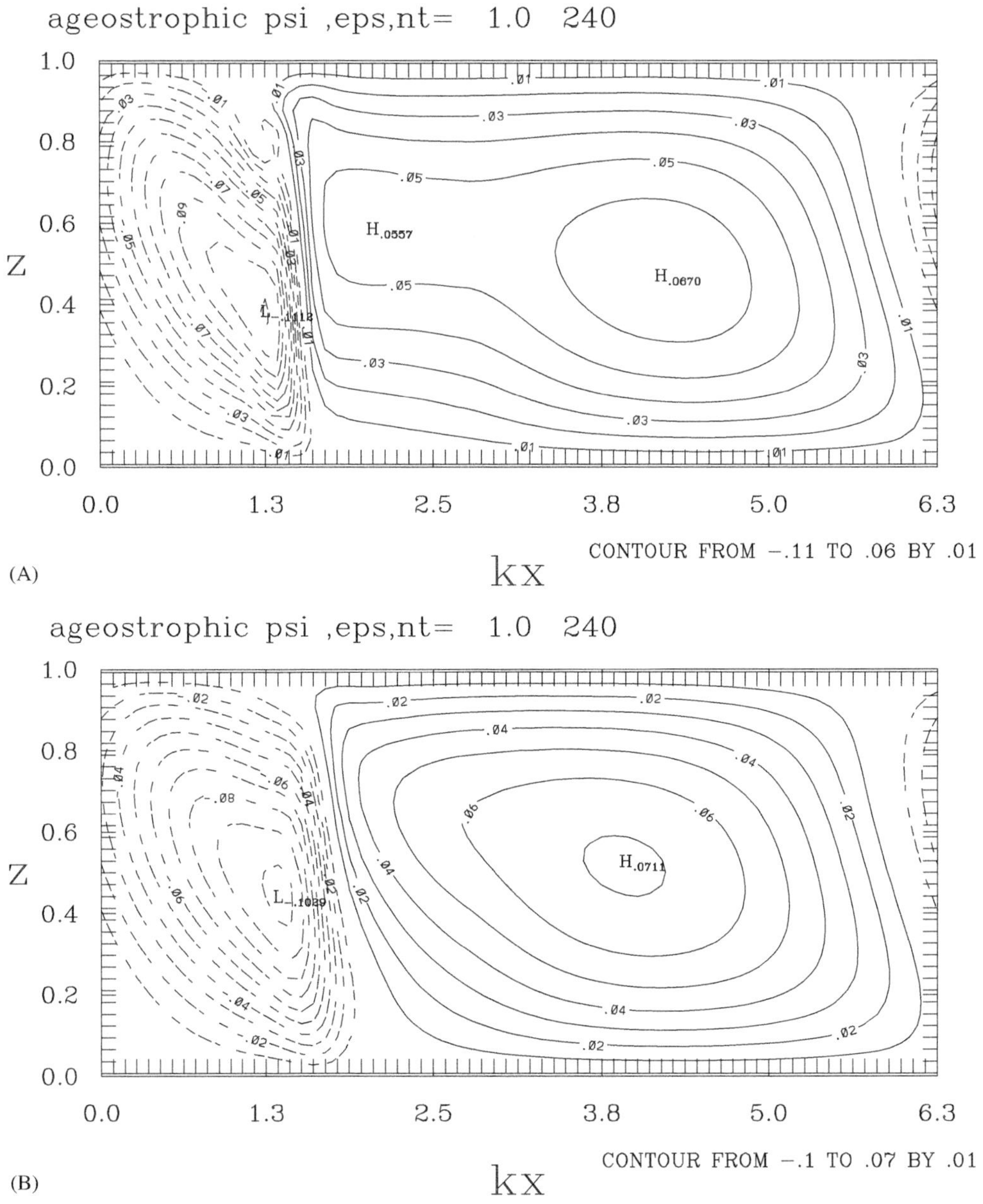

Figure 6.8. Structure of the nondimensional ageostrophic potential field on day 2.5 for (a) profile LOW and (b) profile HIGH in the case of an Eady flow as the reference state.

reference for comparison. The total state in this SG model is again the sum of a reference state and a time-dependent departure state. The former is characterized by a zonal baroclinic jet $\overline{u}(y,z)$, in thermal wind balance with a corresponding potential temperature field $\overline{\theta}(y,z)$ and a uniform background stratification $N = \sqrt{\frac{g}{\theta_{oo}}\frac{d\theta_o}{dz}}$. The departure state is composed of a 3D ageostrophic velocity component $(\hat{u}, \hat{v}, \hat{w})$

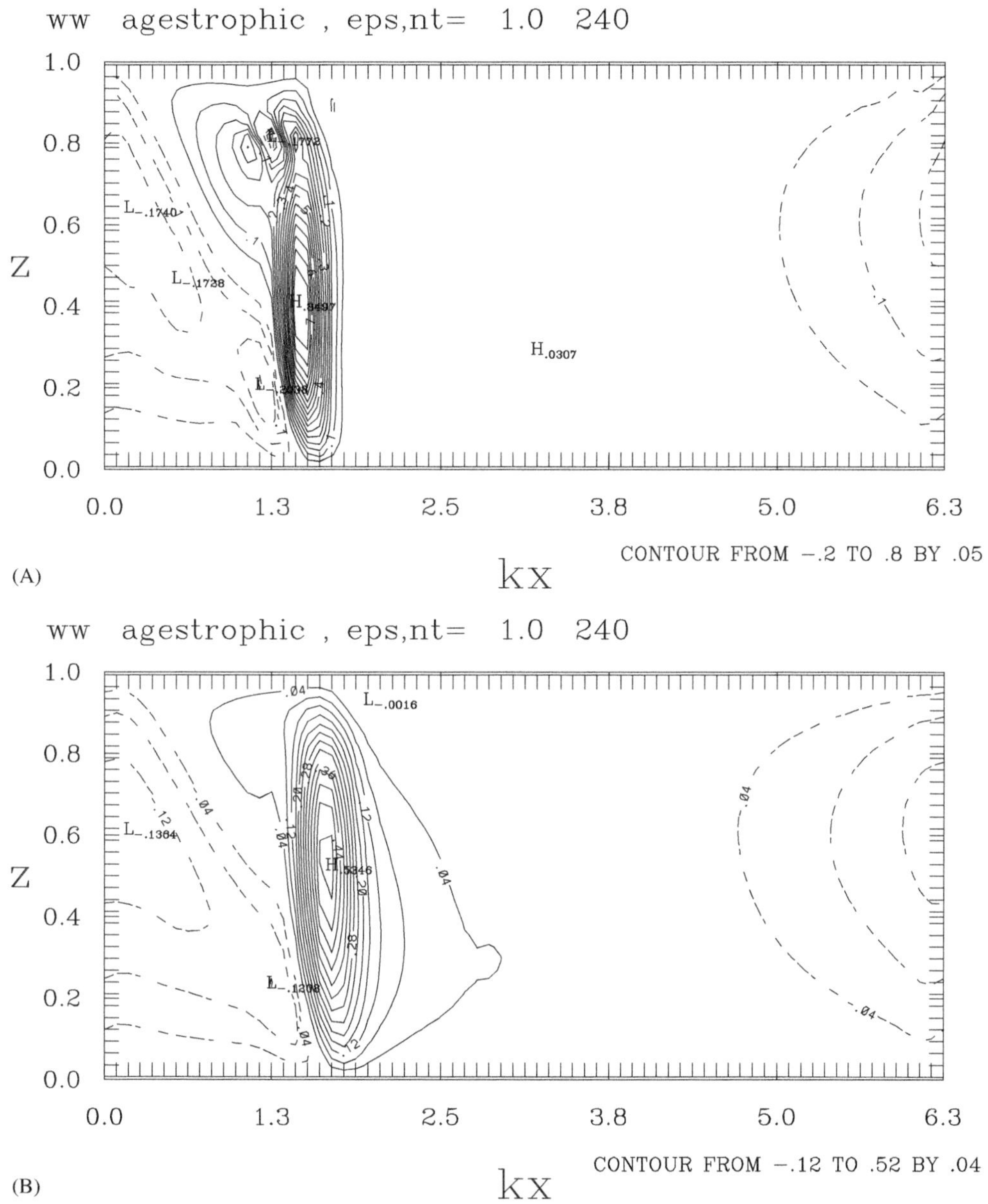

Figure 6.9. Structure of the nondimensional vertical velocity field on day 2.5 for (a) profile LOW and (b) profile HIGH in the case of an Eady flow as the reference state.

and a geostrophic velocity component $(\hat{u}, \hat{v}, 0)$ in thermal wind balance with a component in the potential temperature field $\hat{\theta}$. The set of governing equations for the moist SG system under consideration are stated below for easy reference:

$$0 = -\hat{\varphi}_x + f\hat{v}, \tag{6.12a}$$

$$0 = -\hat{\varphi}_y - f\hat{u}, \tag{6.12b}$$

$$0 = -\hat{\varphi}_z - \frac{g}{\theta_{oo}}\hat{\theta}, \tag{6.12c}$$

$$\tilde{u}_x + \tilde{v}_y + \tilde{w}_z = 0, \tag{6.12d}$$

$$\frac{D(\overline{u} + \hat{u})}{Dt} = f\tilde{v}, \tag{6.12e}$$

$$\frac{D(\hat{v})}{Dt} = -f\tilde{u}, \tag{6.12f}$$

$$\frac{D(\overline{\theta} + \hat{\theta})}{Dt} + \tilde{w}\frac{d\theta_o}{dz} = \alpha h D\{\varsigma^*\}, \tag{6.12g}$$

where

$$\frac{D}{Dt} = \frac{\partial}{\partial t} + u_{total}\frac{\partial}{\partial x} + v_{total}\frac{\partial}{\partial y} + w_{total}\frac{\partial}{\partial z},$$

$$\zeta^* = \nabla^2\psi^*.$$

Apart from the heating term in (6.12g), this set of equations is the same as (2.16). Therefore, incorporating this condensational heating into a 3D SG model only requires a relatively minor extension of its formulation.

The method of analysis of this moist model is the same as that of the counterpart dry SG model. Suffice it to point out here, the differences in the subsequent governing equations. In particular, the SG tendency equation of this moist model, would have one additional term, namely:

$$\begin{aligned} P_t = &-(\overline{u} + \hat{u} + \tilde{u})P_x - (\hat{v} + \tilde{v})P_y - \tilde{w}P_z - \tilde{u}_x\hat{v}_x + \tilde{u}_y\hat{u}_x - \frac{f^2}{N^2}\tilde{u}_z\hat{v}_z \\ &-\tilde{v}_x\hat{v}_y + \tilde{v}_y\hat{u}_y + \frac{f^2}{N^2}\tilde{v}_z(\overline{u}_z + \hat{u}_z) - \tilde{w}_x\hat{v}_z + \tilde{w}_y(\overline{u}_z + \hat{u}_z) - \frac{f^2}{N^2}\tilde{w}_z\psi_{zz}, \\ &+\tilde{v}_y\overline{u}_y + (\hat{v} + \tilde{v})\overline{u}_{yy} + \tilde{w}\overline{u}_{zy} + \frac{fg\alpha}{N^2\theta_{oo}}h_z D\{\varsigma^*\} \end{aligned} \tag{6.13}$$

$$P = \left(\frac{\partial^2}{\partial x^2} + \frac{\partial^2}{\partial y^2} + \frac{f^2}{N^2}\frac{\partial^2}{\partial z^2}\right)\psi, \tag{6.14}$$

where P is identifiable with the quasi-geostrophic potential vorticity. The constraint $\tilde{w} = 0$ at the top and bottom surfaces leads to boundary conditions according to the thermodynamic equation:

$$\begin{aligned} &\text{at } z = 0, \text{D:} \\ &\psi_{zt} = -(\overline{u} + \hat{u} + \tilde{u})\psi_{xz} + (\hat{v} + \tilde{v})(\overline{u}_z + \hat{u}_z). \end{aligned} \tag{6.15}$$

It was shown in Section 2.5.4 that the 3D ageostrophic velocity field of a Boussinesq fluid may be represented by two ageostrophic potential fields, η and χ.

They are related to the 3D geostrophic streamfunction field ψ by two coupled equations referred to as *generalized Sawyer–Eliassen* equations, (2.25) and (2.26). The LHS of the counterparts of (2.25) and (2.26) would remain unchanged. But the RHS of the counterpart of (2.25) and (2.26) would contain an additional term arising from the condensational heating, namely:

$$G\{\psi\} = 2(\bar{u}_z + \hat{u}_z)\hat{v}_x + 2\hat{v}_y\hat{v}_z + \frac{g\alpha}{f\theta_{oo}} h \frac{\partial}{\partial x} D\{\varsigma^*\}, \tag{6.16}$$

$$H\{\psi\} = -2(\bar{u}_z + \hat{u}_z)\hat{u}_x - 2(\bar{u}_y + \hat{u}_y)\hat{v}_z + \frac{g\alpha}{f\theta_{oo}} h \frac{\partial}{\partial y} D\{\varsigma^*\}. \tag{6.17}$$

The algorithm would be to determine the new field of ψ by integrating (6.13), (6.14) and (6.15) for one time-step. It is followed by computing the corresponding η and χ fields in a moist model with (6.16) and (6.17) as the forcing functions. In this way, the integration can proceed to the next time-step and thereafter.

6.4.1 Results

We now make a close comparison of two simulations using the same 3D SG model setting: one without heating and the other with parameterized heating. The initial condition of both runs is the one shown in Figure 5.3. As expected, the initial baroclinic disturbance intensifies at a faster rate under the influence of self-induced heating. It is found that frontal collapse occurs in the model shortly before day 5 when there is heating in contrast to being shortly after day 5 in the case of no heating. The distribution of the heating rate at the mid-level on day 4.875 (468th time-step) is shown in Figure 6.10. The zero contours are suppressed. The parameterized condensational heating occurs only over the areas where the low-level vorticity is cyclonic. The result indicates that there is parameterized condensational heating over two long zones. In light of the surface temperature field (to be shown shortly in Figure 6.13), we see that one area is ahead of a surface cold front and the other area is associated with a warm front that extends to the surface low-pressure center. The disturbance potential temperature field at the mid-level on day 4.875 is shown in Figure 6.11. These two figures jointly suggest a positive correlation between the heating field and the temperature field. The positive correlation between the two patterns in Figures 6.10 and 6.11 implies that the heating contributes to a generation of total disturbance available potential energy. The heating also indirectly affects the energetics, since all field variables are impacted.

By a simple extension of the energetics analysis discussed in Section 5.5, we would get the dimensional equation for the rate of change of the total disturbance energy E with heating as follows:

$$<E>_t = <C(\bar{K}, K)> + <C(\bar{A}, A)> + \frac{\alpha g}{\theta_{oo}\theta_{oz}} <h\theta D\{\zeta^*\}>. \tag{6.18}$$

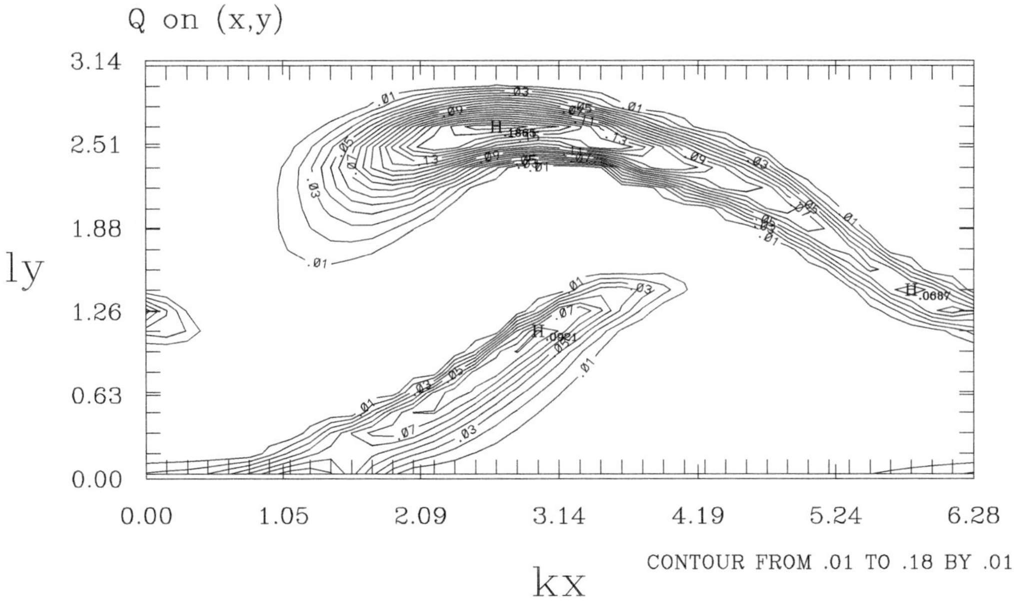

Figure 6.10. Distribution of the nondimensional parameterized condensational heating rate at mid-level on day 4.875 (468th time-step) in the 3D SG model.

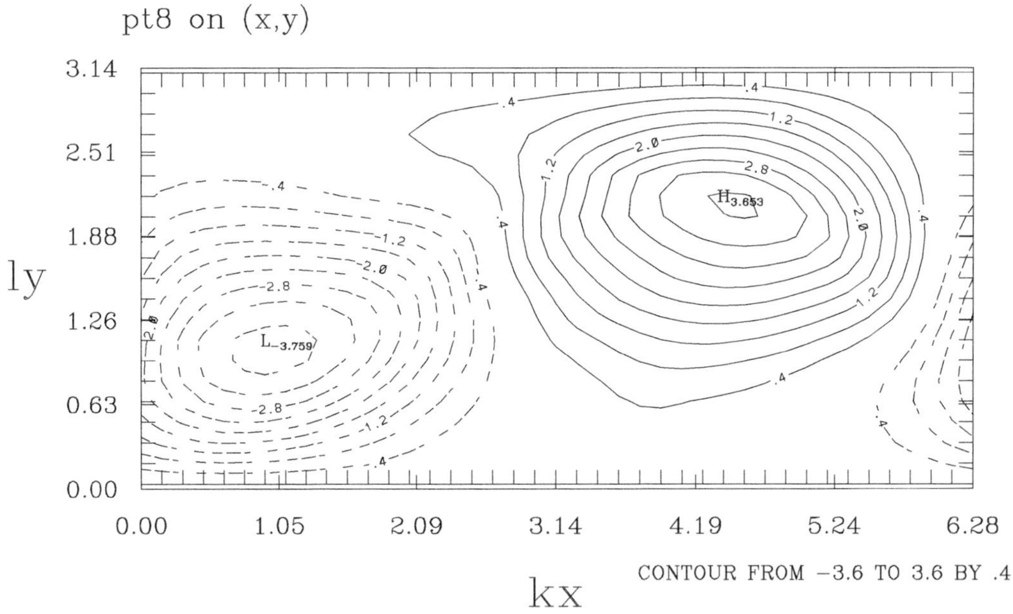

Figure 6.11. Distribution of the disturbance potential temperature at mid-level $z = 0.5$ on day 4.875 (468th time-step).

The angular bracket stands for domain integral. The first two terms on the RHS of (6.18) stand for the generation rate of the domain total disturbance kinetic energy K and available potential energy A by the dynamical processes. They result from conversion of the corresponding energy components of the reference state to those of the

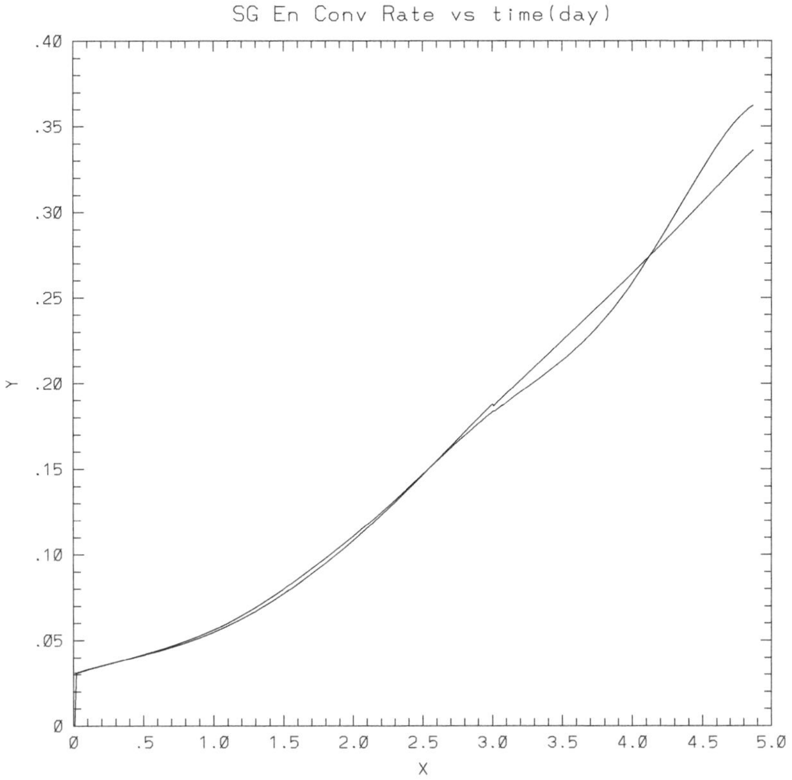

Figure 6.12. Evolution of the generation rate of total disturbance energy in units of $\frac{f\lambda^3 D^2}{N}$ by the dynamical processes alone $\langle C(\overline{K},K)\rangle + \langle C(\overline{A},A)\rangle$ in the two simulations with versus without heating.

departure state. The two terms are measures of the indirect effect of the heating on the energetics. The third term stands for the generation rate of the total available potential energy A by the heating itself. This is a direct effect of the heating from the perspective of energetics. It would be instructive to compare the results of $\langle C(\overline{K},K)\rangle + \langle C(\overline{A},A)\rangle$ in the two simulations with versus without heating. The results are shown in Figure 6.12. The two curves start off with identical values by choice and increase together almost indistinguishably up to about day 3. The value in the simulation with heating is larger than that without heating on day 4.875 by about 10 percent (0.363 versus 0.336). It is interesting to find that the value for the simulation with heating is not always larger at all times throughout the evolution.

The nondimensional form of the generation rate of total available potential energy by the heating is $\Gamma\langle\tilde{\theta}\tilde{Q}\rangle$ where $\Gamma = \frac{g\Theta}{\lambda N D \theta_{oo}}$ and $\tilde{Q} = \tilde{\alpha} h D\{\tilde{\zeta}^*\}$. This quantity is found to have a value of 0.043 on day 4.875, which amounts to about 12 percent of the generation rate by the dynamical processes at that time. In other words, the total increase by the heating on the generation rate of the total disturbance energy is over 22 percent. This is a substantial impact on the frontal development, although fronts would form whether or not there is heating in this setting.

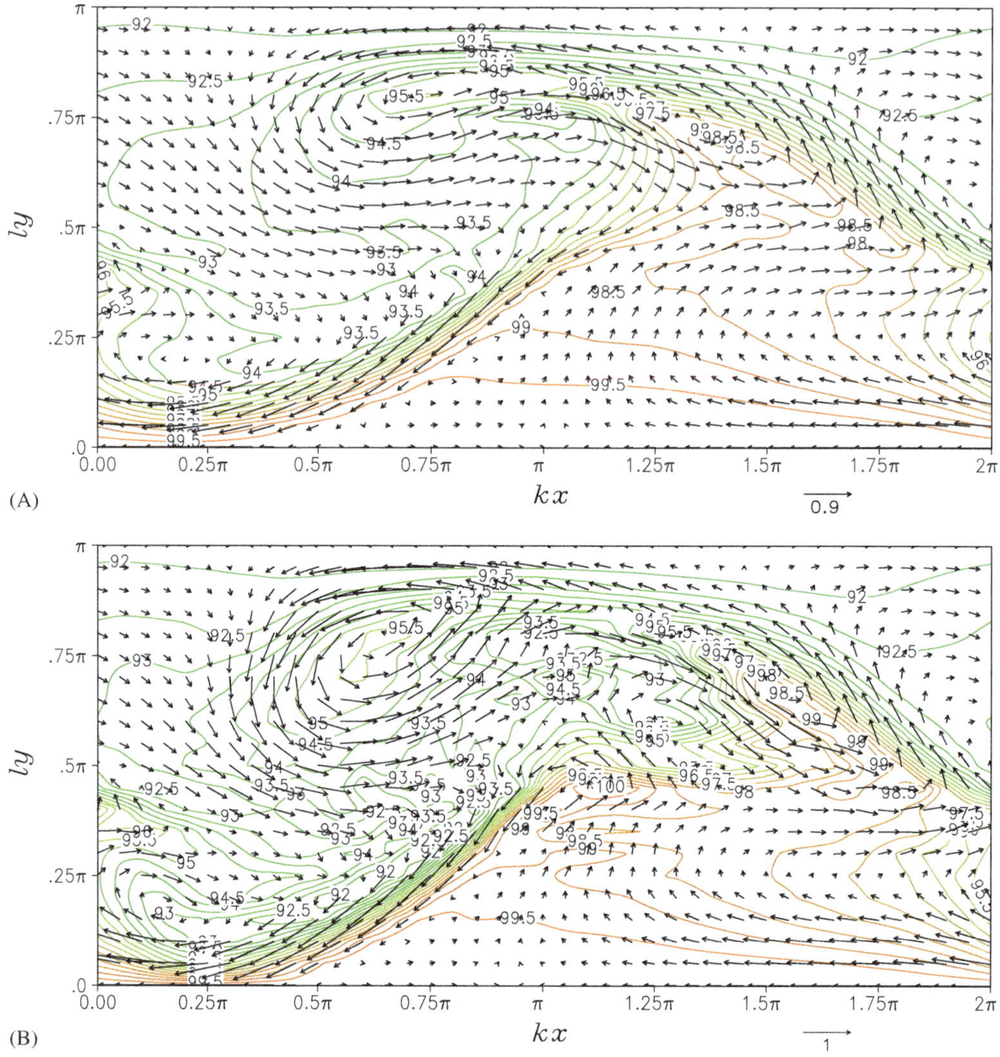

Figure 6.13. Distribution of nondimensional geostrophic velocity $(\hat{u}, \hat{v})$ vector field and potential temperature $(\overline{\theta}+\hat{\theta})$ field in units of $\Theta = 3\,\mathrm{K}$ at the bottom surface $z=0$ on day 4.875 (a) without heating and (b) with heating.

The nondimensional geostrophic velocity vector field together with the total potential temperature field at the bottom surface $z=0$ at this time are compared in Figure 6.13a,b. It is seen that the geostrophic velocity component at both fronts is stronger by 20 percent with the heating. The fronts in panel (b) are distinctly sharper than those in panel (a). But the broad similarity of the two panels again confirms that while the heating process enhances the frontal development, it is not a decisive factor.

The corresponding plots of the ageostrophic velocity vector field with the potential temperature field at $z = 0$ on day 4.875 are compared in Figure 6.14a,b. The ageostrophic velocity field is almost perpendicular to the cold front with or without heating.

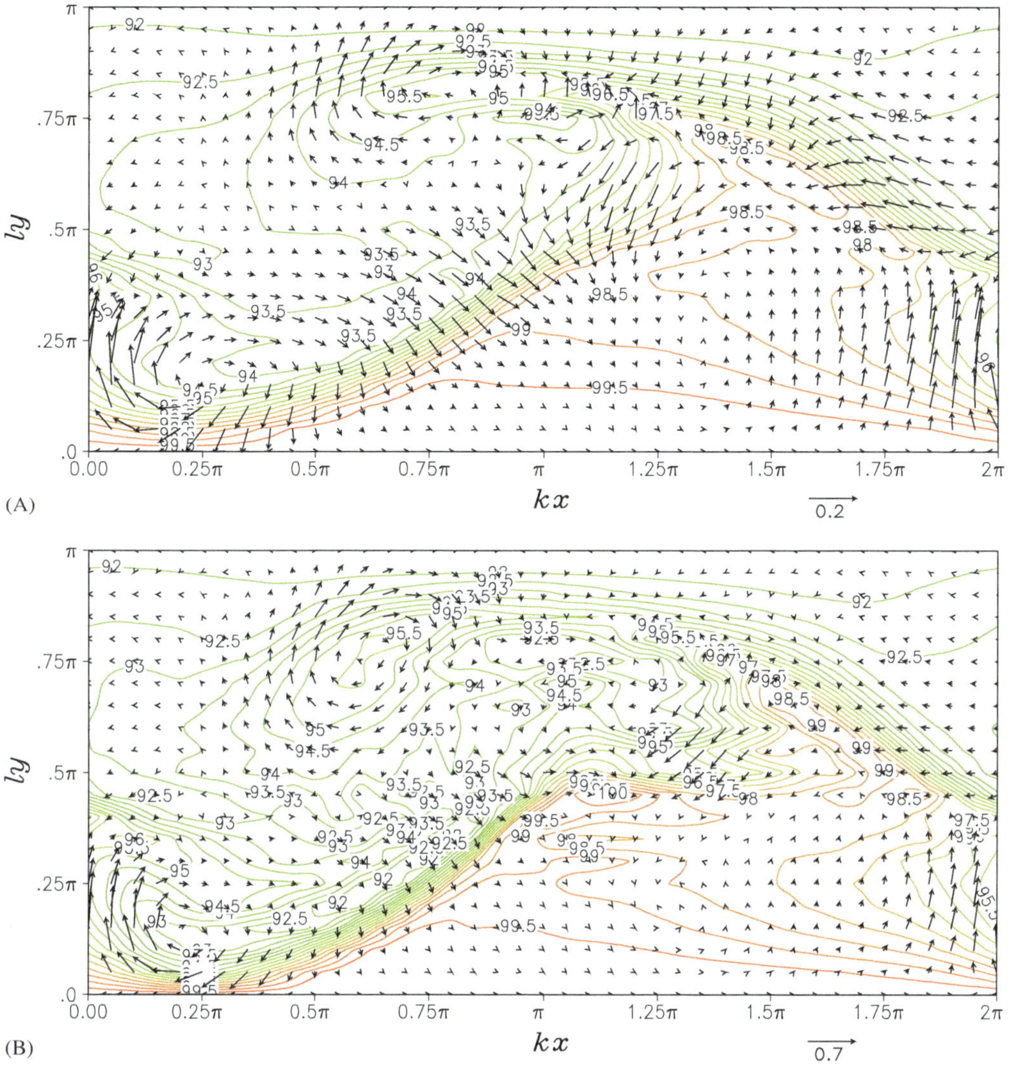

Figure 6.14. Distribution of ageostrophic velocity $(\tilde{u}, \tilde{v})$ vector field in units of λD and potential temperature $(\overline{\theta} + \hat{\theta})$ field in units of $\Theta = 3\,\mathrm{K}$ at the bottom surface $z = 0$ on day 4.875.

When there is heating, the maximum value of ageostrophic velocity is more than three times stronger than when there is no heating (maximum value 0.7 versus 0.2).

Figure 6.15a,b is a comparison of the distribution of the vertical velocity just above the implicit moist layer (at $z = 0.15$) and the geostrophic velocity field at $z = 0$ on day 4.875 with versus without heating. We see ascending motion to the east of the surface cold front and descending motion behind the cold front. The values of ascent are generally twice as large with heating as without heating. This is a much smaller ratio than that in the 2D model simulations. There is also ascending motion over the warm front and descending motion to the west of it.

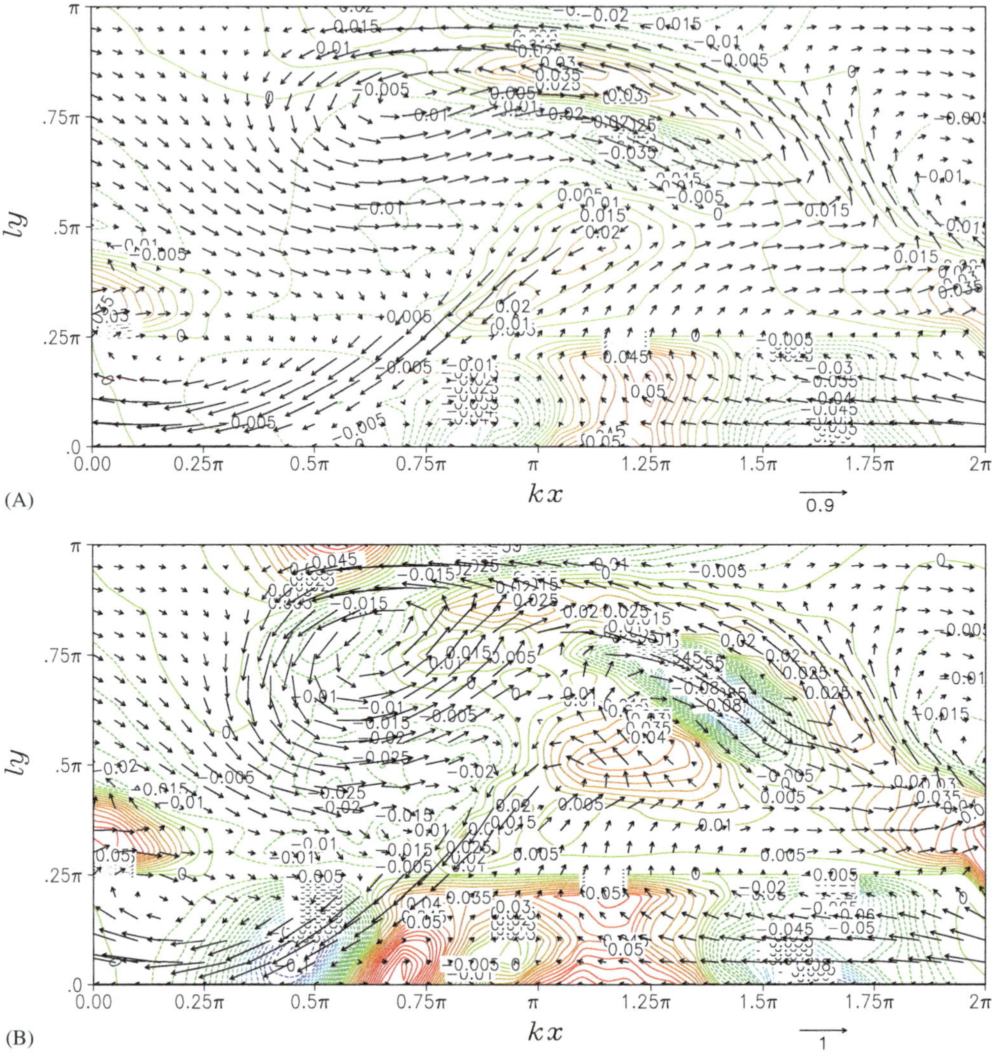

Figure 6.15. Distribution of vertical velocity (contours) at z = 0.15 in units of $\frac{f\lambda D}{N}$ and geostrophic velocity vector field in units of λD at the bottom surface $z = 0$ on day 4.875 (a) without heating and (b) with heating.

Recap

- One simple convection parameterization scheme invokes the notion of "effective stratification" (smaller in the region of ascent than in the region of descent); another simple scheme assumes frictionally induced moisture convergence from a surface moisture layer (this scheme is applied in the analyses of this chapter). It is important for the heating to be positive-definite in either scheme so as to allow for an intrinsically nonlinear response.
- Heating in a 2D SG frontal Eady model significantly enhances intensification of an unstable disturbance, including its ageostrophic circulation in the cross-section normal to the model fronts.
- Advective effect of such ageostrophic flow enhances frontogenetic development with significant impacts upon the structure of all properties of the disturbance.
- There is a stronger impact of heating on the bottom surface front than on that at the top surface because the heating below the level of maximum heating increases the potential vorticity in the lower troposphere and thereby a greater surface frontogenetic development.
- The impact of latent heat release on model cold front leads to an earlier frontal collapse.
- Results of the 3D SG moist frontal model analysis corroborate and generalize those of the counterpart 2D model.
- The dynamical difference between cold front and warm front development in both dry and moist 3D SG models partly stems from the fact that the thermal advection by the geostrophic and the ageostrophic velocity components reinforce one another in the cold front region, but are of opposite signs in the warm front region.

7

General Dynamics of Fronts

7.1 Introductory Remarks

In this chapter we investigate the general dynamics of frontogenesis with a high-resolution non-hydrostatic **A**dvanced **R**esearch **W**eather **R**esearch and **F**orecasting (ARW-WRF) model including frictional and diabatic processes. While it is a comprehensive model, a realistic baroclinic jet with a stratospheric layer similar to the observed winter zonal mean state in the mid-latitudes is introduced at the initial time as a form of dynamical forcing. The technical aspects of the model will be described in Section 7.2. This model is dynamically much less constrained than the semi-geostrophic (SG) models used in the previous chapters. We waited until this chapter to analyze the dynamics of upper-level fronts (ULFs) for quantitative discussion, although an example of a ULF in a cyclone–front system was shown for illustration in Section 1.1. The ARW-WRF model enables us to address a number of fundamental issues concerning both the surface fronts (SFs) and ULFs. A broad comparison of the simulations with versus without diabatic heating will be first made in Section 7.3. Suffice it to mention for the moment that there is good reason to measure the rates of development of SF and ULF differently. The nature of this important issue will be elaborated in Section 7.5. The explicit form of the frontogenetical function appropriate for SF, $F_2 \equiv \frac{D}{Dt}|\nabla_2\theta|$, has been derived in Section 1.3, whereas that for ULF, $F_3 \equiv \frac{D}{Dt}|\nabla\theta|$, will be derived in Section 7.6.1. They will be applied in Section 7.4 and Section 7.6 respectively. Most of the figures used in our discussions are taken from two recent articles (Mak et al., 2016, 2018), referred to as **MLD1** and **MLD2** for short.

In Section 7.4, we will specifically investigate *why cold fronts are characteristically much narrower, longer, and more intense than warm fronts.* It is not obvious as to why that should be so. Hoskins and Heckley (1981) conclude in their 3D SG model simulation study that "many of the distinctions between cold fronts and warm fronts are caused by the dynamical structure of growing baroclinic waves associated with the forward tilt of the temperature wave." While this statement is fine in itself, it is not sufficiently specific because all synoptic-scale unstable waves of a baroclinic shear flow must have a forward tilt in their temperature field as required by the consideration of energetics even in the context of linear quasi-geostrophic (QG) baroclinic instability. The many studies about the effects of friction on fronts are reviewed in MLD2. It may be broadly stated that the impacts of friction per se on SFs can be significant by limiting their minimum width and maximum intensity. There is considerably more uncertainty in how we might incorporate the diabatic heating in a model. Its impact upon frontogenesis has received, not surprisingly, less attention. It is imperative to closely ascertain the role of diabatic heating in the distinct differences between cold front and warm front.

In Section 7.5, we will investigate the dynamics of upper-level frontogenesis. A literature review concerning upper-level fronts was given in Section 1.4. On the basis of the WRF model simulation we will ascertain the horizontal and vertical structure of ULFs and determine the roles of the transverse circulation and diabatic heating. We will establish the relative importance of the different processes in their contributions to the constituent parts of the 3D frontogenetic function F_3.

In Section 7.6 we will address the issue of whether or not an SF would synergistically interact with a ULF: If so, under what circumstances, how, and why? This is a dynamic problem of theoretical and practical importance. A literature review of the related studies has been given in Section 1.4. We will quantify how the observed ULF and the SF shown in Figure 1.1 interact. We first give a preliminary discussion of the dynamic nature of such interaction. Then we report a complementary diagnosis of the interaction of ULF with the cold SF in our moist model simulation. It helps clarify the dynamic nature of such an interaction.

This chapter ends with some concluding remarks in Section 7.7.

7.2 A State-of-the-Art Model for Frontal Simulation

The WRF model core is based on a Eulerian solver for the fully compressible non-hydrostatic equations, cast in flux (conservative) form in a terrain-following eta-coordinate. Prognostic variables for this solver are column mass of dry air (mu), velocities u, v and w (vertical velocity), potential temperature, and geopotential. Non-conserved variables (e.g., temperature, pressure, density) are diagnosed from the conserved prognostic variables. The solver uses a third-order Runge–Kutta time integration scheme coupled with a split-explicit second-order time integration scheme for the acoustic and gravity wave modes. Fifth-order upwind-biased advection operators are used in the fully conservative flux divergence integration; Second- to sixth-order schemes are run-time selectable. It conserves all the fundamental properties such as total mass, energy, and potential vorticity. We use a model horizontal domain 4005 × 8010 km depicted by 263 × 534 grid points so that the grid resolution is $dx = dy = 15$ km. The top of the model is at about $z = 16$ km, corresponding to about 150 hPa. The vertical domain has 64 slightly stretched levels

($dz \approx 250$ m). The horizontal grid arrangement is Arakawa C-grid staggering. The time-step is 90 s with 22.5 s for acoustic waves. The east–west boundary conditions are periodic. The north–south boundary conditions are symmetric. The bottom boundary is a flat surface through which there are momentum, sensible heat, and moisture transfers. The top boundary is a constant pressure surface. No damping is applied near the top of the domain. Full Coriolis terms are included. There are various physics packages to choose from for depicting the subgrid processes and atmospheric radiation. We make two simulations and diagnose the model output in the context of frontal dynamics. All diabatic heating processes are switched off in one simulation, except for the boundary layer dissipation. It is referred to as the "Dry Run." The other simulation incorporates all diabatic heating processes and is referred to as the "Control Run." The model physics schemes in this run are switched on with the use of the following parameters: (a) mp_physics = 10, Morrison et al. two-moment scheme; (b) cu_physics = 14, new GFS simplified Arakawa–Schubert scheme from YSU; (c) bl_pbl_physics = 1, Yonsei University PBL (YSU) scheme; (d) sf_sfclay_physics = 1; MM5 Monin–Obukhov scheme (similarity theory); and (e) ra_lw_physics = 4, rapid radiative transfer model (RRTMG) long-wave scheme and ra_sw physics = 4, for short-wave radiation.

The details of the physics schemes incorporated in the model are elaborated in the user manual (see www2.mmm.ucaar.edu/wrf/users). It goes without saying that no individual researcher can be expected to be fully knowledgeable of all technical details in such a wide assortment of schemes. An application using such a model therefore rests upon the fruit of the collective efforts of the atmospheric science community, with the price of accepting various inherent limitations.

7.2.1 Initial Condition

The initial state of our model atmosphere is constructed in two steps. The first step is to construct a zonally symmetric component with the code of 2D potential vorticity (PV) inversion used by Plougonven and Snyder (2007). The code is based on an over-relaxation method. The PV value in the model troposphere and stratosphere is first assigned 0.4 and 5.0 PVU respectively ($1\ \text{PVU} = 10^{-6}\ \text{m}^2\ \text{s}^{-1}\ \text{K(kg)}^{-1}$). A built-in code with a number of adjustable parameters enables us to further prescribe the structure, location, and strength of a baroclinic jet with a sloping tropopause. The algorithm of PV inversion is relatively slow to converge because the matrix associated with the high resolution of the model is very large. When the iteration procedure is eventually stopped, the change of the field from the last iteration is deemed acceptably small although there remain small variations in the PV field within the model troposphere and stratosphere. At this point, we apply the kick-off procedure in the WRF source code designed for the specific purpose of simulating 3D baroclinic waves. It uses a thermal bubble (perturbation θ with maximum of 1 K) at the center of the domain. The characteristic length scale in x-, y- and

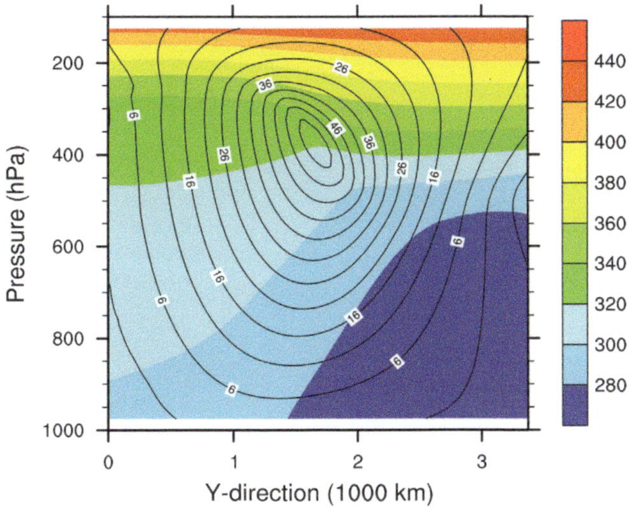

Figure 7.1. Structure of the reference state in terms of a zonally uniform zonal velocity in m s^{-1} (contour) and potential temperature in K (shading) (taken from MLD1).

z-directions of the bubble is 4000 km, 2000 km and 8 km, respectively. We reduce the initial noisy oscillations arising from the imbalance of such initial condition by integrating the model for an inertial period (37 hours in this study) and time-averaging all fields. This procedure is repeated once more to obtain an internally consistent balance initial state. Time is then reset to zero when we start the actual simulations. The resulting initial state of the model is referred to as the *reference state*, with an average surface pressure of 1014 hPa. The zonal mean component of such a state is shown in Figure 7.1. It is characterized by a zonally uniform baroclinic jet with a maximum speed of 60 m s^{-1} centering at about 350 hPa over the center of the meridional domain. The potential temperature field has a stable stratification everywhere, with much larger values above the level of a model tropopause.

The corresponding distributions of the potential vorticity and potential temperature of this reference state are shown in Figure 7.2. The 1.5 PVU and 3.0 PVU contours are highlighted. The former may be regarded as a meaningful indicator of the tropopause. The initial model tropopause has realistic features: being quite high in the southern portion of the domain at about 275 mb, sharply decreasing through a middle region and tapering off to a considerably lower level (~500 mb) in the northern portion of the domain. The baroclinic jet is centered at the middle of the tropopause break. We have not fine-tuned the reference state. The broad characteristics of the evolution of a flow in this model are not sensitive to the details of the reference state. The slight variations in the PV field within the model troposphere and stratosphere stem from the less than perfect convergence of the PV inversion algorithm.

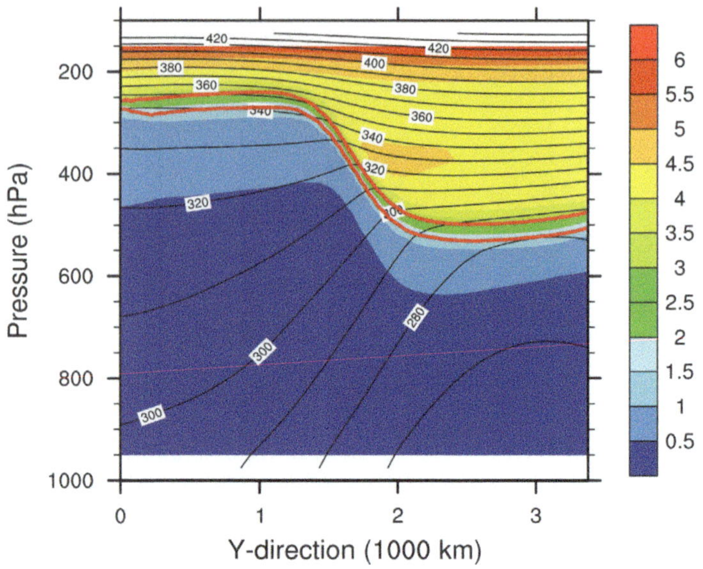

Figure 7.2. Distributions of potential vorticity in PVU (shading) and potential temperature in K (contours) of the reference state; red lines highlight the contours of 1.5 PVU and 3.0 PVU (taken from MLD1).

For the Control Run that incorporates the diabatic heating processes, we introduce the following distribution of moisture at $t = 0$ as a part of the initial state. The initial distribution of the mixing ratio is zonally uniform with a surface value of 16 g kg^{-1} decreasing northward to 1 g kg^{-1} from the southern to the northern boundary of the domain. Vertically it is mostly confined to the lower troposphere.

7.3 Overall Evolution of the Flow in the Dry Run and Control Run

The disturbance field quickly intensifies and the main feature is an unstable baroclinic wave. As expected, SFs and ULF form in a few days as a consequence of the nonlinear dynamics of the wave when it has become sufficiently strong. A feel for the overall development of the flow can first be gained by examining how the minimum surface pressure of the baroclinic wave evolves (Figure 7.3). It is found that the evolution in the first 4.5 days in both runs is essentially identical because the diabatic processes are too weak to make a measurable impact up to this point. The minimum pressure in the Dry Run evolves toward a value of about 980 hPa, whereas that in the Control Run asymptotes toward about 970 hPa. In other words, the surface pressure of the disturbance without diabatic heating has deepened by about 35 hPa in 6.5 days. The one with diabatic heating has deepened by about 45 hPa in 7 days.

We now proceed to diagnose the model output for the purpose of determining the 3D structures of the SFs and the ULF and delineating their dynamical nature. We will first discuss surface frontogenesis in Section 7.4 and then upper-level frontogenesis in Section 7.5.

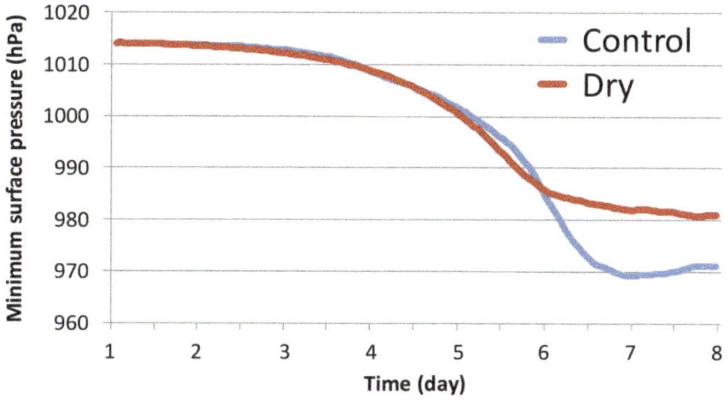

Figure 7.3. Evolution of the minimum surface pressure in hPa of the disturbance in the Dry Run and the Control Run (taken from MLD1).

7.4 Surface Frontogenesis

The discussions here focus on the characteristics and physical nature of surface frontogenesis as deduced from the model simulations. Since the bottom surface of the model is homogeneous and the east–west boundary conditions are cyclical, there is no particular significance in the location of the origin of the x-coordinate. Thus, we are free to shift the location of $x = 0$ when we graphically present the results. It has been shifted eastward by 230 grid points in the figures of this section so that the key features would appear near the center of the plotting domain.

7.4.1 Rudimentary Comparison of the SFs With Versus Without Diabatic Heating

We begin by taking a close look at the flow and temperature fields in the frontal regions. Panels (a) and (b) of Figure 7.4 give a comparison of the surface pressure and temperature of the disturbance in the Dry Run and Control Run on day 6.5. The minimum surface pressure of the disturbance in the Control Run is lower than that in the Dry Run, about 972 versus 988 hPa. The baroclinic wave has propagated further to the east in the Control Run than in the Dry Run, suggesting that the steering level is higher under the influence of diabatic heating. Panel (c) highlights the difference of the disturbance in the two runs by showing their difference in the surface pressure distribution. It is quite evident in these plots that there is an intense surface cold front as well as a warm front embedded in the disturbance with or without diabatic heating. The fronts are more pronounced in the Control Run. If the Control Run were a proxy of a real disturbance and the Dry Run were used as a basis for making a forecast, the disturbance's intensity would be underestimated in the forecast. The surface pressure to the east of the disturbance center would be significantly overestimated and that to the west underestimated. There would be corresponding discrepancies in the wind field and weather elements in the forecast.

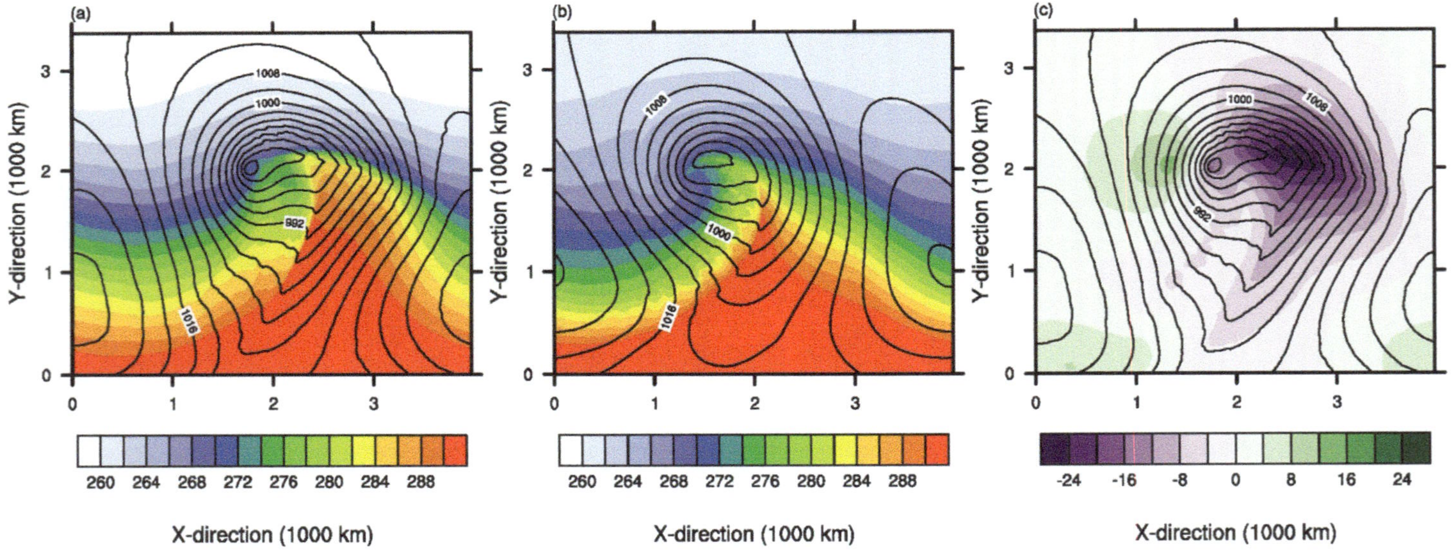

Figure 7.4. Distribution of surface pressure (contours) in hPa and surface temperature (shading) in K of the (a) Control Run and (b) Dry Run on day 6.5. Panel (c) again shows the surface pressure in Control Run (contours) and the difference (Control – Dry) (shading).

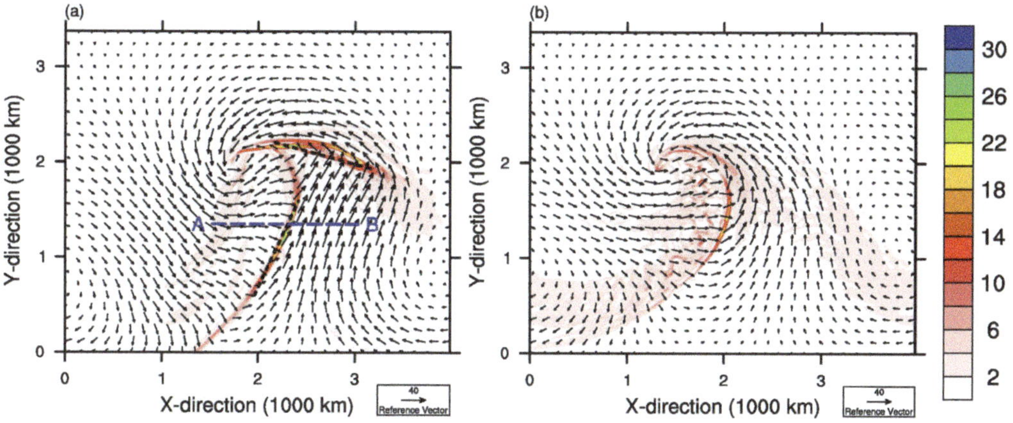

Figure 7.5. Distributions of the wind (vectors) and $|\nabla_2\theta|$in K per 100 km (shading) on 950 mb in the (a) Control Run and (b) Dry Run on day 6.5. Line of AB in panel (a) is the location for examining the vertical structure of the thermal field in Figure 7.7.

To quantitatively ascertain the differences in the structural properties of the surface cold and warm fronts with versus without diabatic heating, we examine the distributions of $|\nabla_2\theta|$ where θ is the potential temperature on the 950 hPa surface on day 6.5 of both runs. The result is shown in Figure 7.5 together with the corresponding wind field. The cold front is distinctly longer, narrower, and more intense than the warm front, even without diabatic heating. The cold front is more separated from the warm front when there is diabatic heating. The intensity of the cold front with diabatic heating is distinctly more intense than that without heating. The former reaches about 20 K per 100 km whereas the latter is about 12 K per 100 km, as seen in Figure 7.5a,b. In each case, the warm front is considerably broader than the corresponding cold front. The general distribution of the surface wind is quite realistic, especially in the case of diabatic heating with a signature southerly in the warm sector. It is also interesting to find a trace of a secondary cold front further to the west of the primary cold front in the Control Run, sometimes observed in the atmosphere. We will shortly examine the vertical thermal structure over the cold front on an east–west oriented cross-section along the line AB.

7.4.2 Movement, Structure, and Evolution of the SFs in the Control Run

In Section 1.5 we derived the mathematical expressions for a frontogenetical function, $F_2 \equiv \frac{D}{Dt}|\nabla_2\theta|$. It is a measure of the material rate of change of the magnitude of the horizontal gradient of potential temperature. This definition is referred to as the "frontogenetical function on x–y plane" for short. This is used in the following analysis of the dynamic nature of surface frontogenesis.

We have constructed a video clip that depicts the simultaneous evolution of the potential temperature and F_2 fields at 925 hPa from day 5 to day 8 in the Control Run.

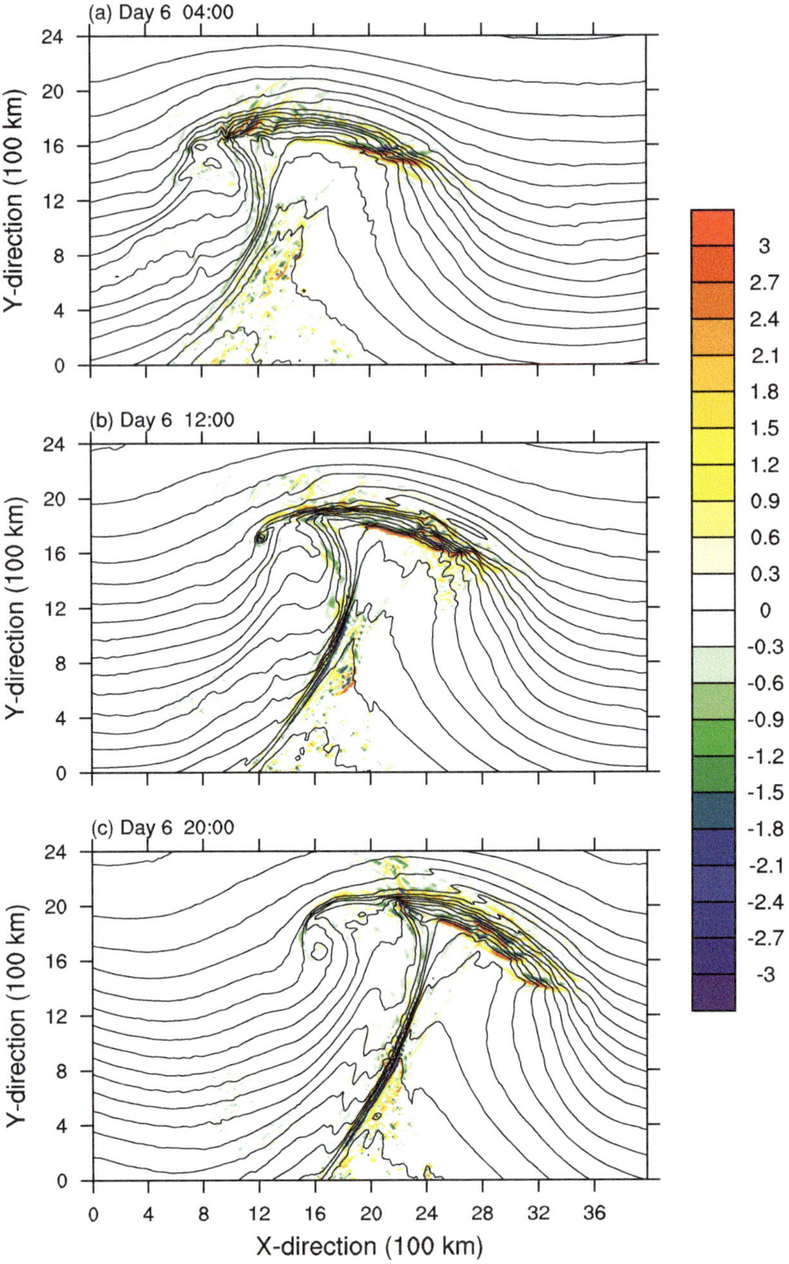

Figure 7.6. Evolution of the surface fronts in terms of potential temperature (contours, CI = 2 K) and F_2 (shading) at 925 hPa in 10^{-8} K m s^{-1} in the Control Run from the 4th to the 20th hour of day 6 in the Control Run (taken from MLD2).

The animation enables us to closely examine how the model SFs evolve in the mature stage of development, thereby to quantitatively identify the distinct differences between the cold and warm fronts. Suffice it to select and show in Figure 7.6 three frames of the animation, taken eight hours apart. The contour interval of potential temperature is 2 K and F_2 is in units of 10^{-8}K m s^{-1}. A *positive value of F_2 means a local frontogenetical effect and negative value means a local frontolytic effect.* The distribution of F_2 in each panel matches well with the locations of the fronts, confirming that it is an appropriate metric for measuring their rate of development.

The three panels of Figure 7.6 vividly reveal the evolution of the frontal system from the 4th to 20th hour of day 6. The cold front is oriented roughly in the SW–NE direction and the warm front largely in the zonal direction. The cold front strongly intensifies and progressively becomes longer and narrower. In contrast, the warm front is distinctly wider, shorter, and weaker, with much smaller changes. While the potential temperature varies by about 14 K over a distance of less than 100 km across the cold front by the 20th hour, the main part of the warm front has such temperature difference over a distance of 250 km. The cold front is about 2000 km long, which is about the length of the intensifying baroclinic wave (of the order of the radius of deformation). On the other hand, the length of the warm front is only about half that.

The cyclone–frontal system as a whole propagates eastward under the influence of a steering flow in the reference state. In addition, the cold front moves faster eastward than the warm front. This relative movement of the cold front is partly due to the cold/dense air wedging under the warmer air ahead and thereby helps push such a front more rapidly eastward. This is analogous of a density current. In contrast, the warm/light air south of the warm front slides and rises over the zonally oriented warm front. While the cold front is displaced eastward by about 1200 km, the warm front only moves by about 800 km in this 16-hour interval, amounting to about 20 m s^{-1} versus 10 m s^{-1}. In other words, the faster movement of the cold front steadily outpaces the warm front. For instance, the northern end of the cold front meets the warm front near its western end on the 4th hour. By the 20th hour, the cold front meets the warm front at its mid-point. The western portion of the warm front, progressively occluded leaves its eastern portion an ever-shortening warm front. All in all, the model simulates quite well all the commonly observed distinct differences between the cold and warm front.

By the 20th hour of day 6, the whole frontal system has the appearance of a jellyfish, with those isotherms along the cold front bunching up together as if they were a single long tentacle. Positive value of F_2 would mean a local frontogenetical effect and a negative value a local frontolytic effect. Movement of a front would manifest as tightly packed bands of F_2 values in alternating signs. If one enlarges panel (c) on a computer screen, one would see that there are tightly packed, long, alternating narrow red and blue bands of F_2 values parallel to the main part of the "tentacle," signifying that the cold front is intensifying and propagating. The F_2 values in the plots quantify the local impact of the physical processes at different parts of the two fronts during their development. For example, there are several pairs of shorter and broader meso-scale green and red bands in the warm front region. Such features suggest that convection has a weaker organization in the warm front region than in the cold front region. The physical nature of these complex features will be clearer when we make close diagnoses of the various constituent parts of F_2 in Section 7.5.6.

7.4.3 Structure of Vertical Velocity

Another interesting characteristic of the frontal circulation is the distribution of the vertical velocity field. The patterns of the vertical velocity field at 900, 700 and 500 hPa on day 6.5 are broadly similar, suggesting a high degree of vertical coherence in its structure. We present the distribution of the vertical velocity on the 900 hPa surface in Figure 7.7a. The ascending motion at this lower tropospheric level near the surface is particularly concentrated along the cold front, with a band of distinct descent a short distance to the west of it. We also show in panels (b) and (c) the vertical distributions of potential temperature and vertical velocity in two cross-sections, one normal to the cold front and the other normal to the warm front. These cross-sections pass through the lines AB and CD respectively. At these locations the vertical velocity is strongest over each front at the top of the corresponding frontal layer. Panel (b) shows that the ascent at the leading edge of the cold front is strongest, reaching 40 cm s^{-1} at about 900 hPa in this cross-section. The ascent reaches the mid-tropospheric levels. The cold front is very narrow and the cold front layer has nearly neutral stratification, undoubtedly resulting from turbulent vertical mixing. Along the top of this cold front layer there are clear indications of a sequence of ascent and descent extending into the interior of the model atmosphere. These are parts of the meso-scale internal gravity wave disturbances elaborated later in Section 7.5. Panel (c) reveals a shallower and wider warm frontal layer that still has stable stratification. It means there is no strong turbulent mixing there. There is ascent broadly distributed over the warm front layer, only about half as strong as that over the cold front. There is no indication of gravity wave excitation over the warm front.

7.4.4 Geostrophic and Ageostrophic Wind Components and Their Thermal Advection

To ascertain the specific cause behind the distinct structural difference between cold front and warm front, we next decompose the surface wind into a geostrophic component and an ageostrosphic component. To simplify the notations, we reserve the subscripts exclusively for denoting partial derivatives. We do not use the commonly used subscripts "g" and "a" to indicate geostrophic and ageostrophic wind components. Instead, we use superscripts "^" and "$\sim$" to indicate the geostrophic and ageostrophic components respectively. Figure 7.8 shows these two velocity components at 950 mb in the Control Run on day 6.5, when the fronts are still rapidly intensifying. As expected, the geostrophic velocity is cyclonic and is generally much stronger than the ageostrophic velocity, particularly in the warm sector ahead of the cold front. In contrast, the ageostrophic wind is in a generally anticyclonic direction. This simply stems from the fact that the wind in a low-pressure region is necessarily sub-geostrophic.

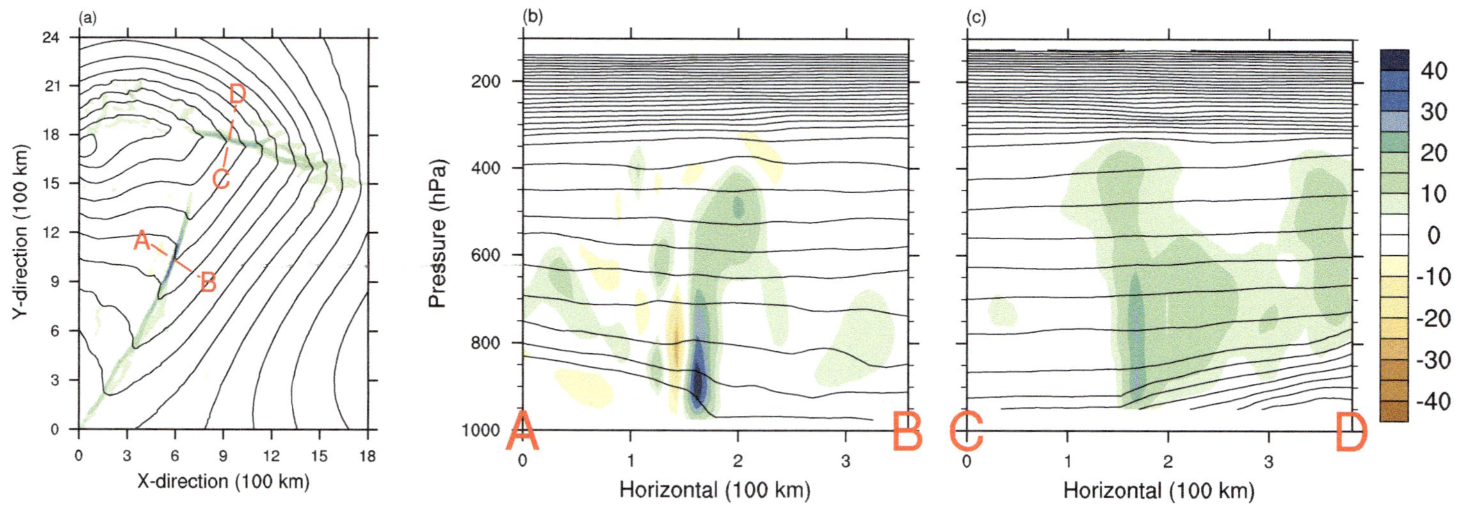

Figure 7.7. (a) Horizontal distribution of the potential temperature (contours) in K and the vertical velocity in cm s^{-1} at 925 hPa. (b) Vertical distribution of potential temperature in K (contours) and vertical velocity (shading) on a cross-section AB normal to the cold front, and (c) on a cross-section CD normal to the warm front in the Control Run on day 6.5.

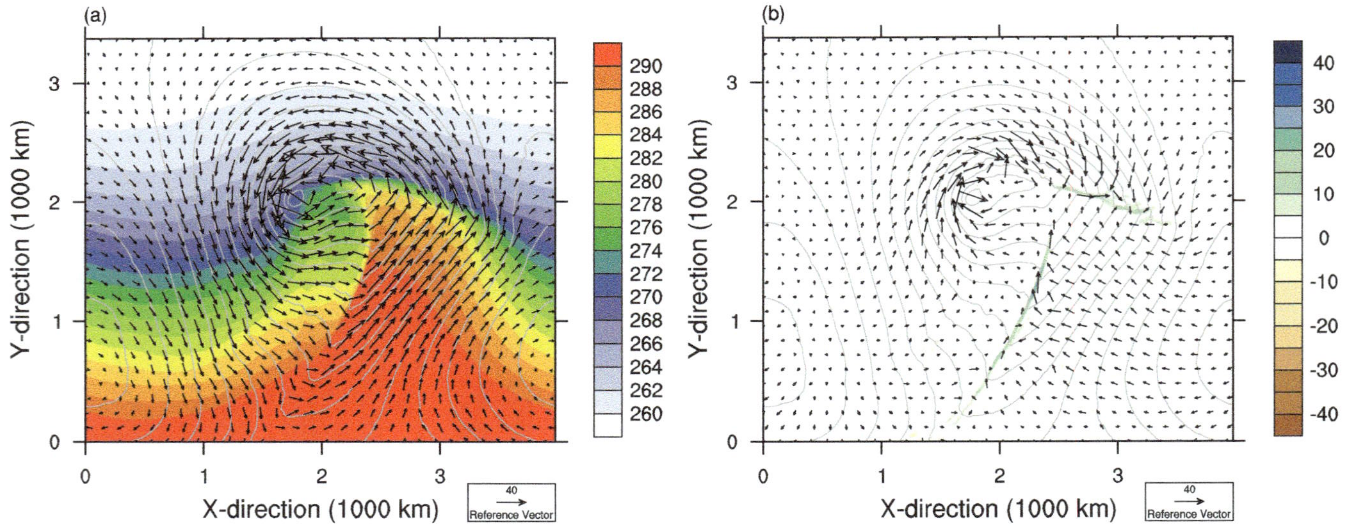

Figure 7.8. Distributions of the (a) geostrophic velocity (vector) and potential temperature (shading) in K and surface pressure (gray contours, min = 974 hPa, CI = 4 hPa), and (b) ageostrophic velocity (vector) and vertical velocity in cm s^{-1} (shading) and surface pressure (black contours) at 950 hPa on day 6.5 of the Control Run (taken from MLD2).

That requires the ageostrophic wind to be in an opposing direction relative to the geostrophic wind. We see in Figure 7.8 that the ageostrophic velocity component is quite substantial near the low-pressure center where the curvature is large.

The difference in the direction of the geostrophic $\hat{v}$ and ageostrophic $\tilde{v}$ wind components in a low-pressure region has a significant consequence for their respective thermal advection, $-\hat{\boldsymbol{v}} \cdot \nabla_2\theta$ and $-\tilde{\boldsymbol{v}} \cdot \nabla_2\theta$. Figure 7.9 shows the distributions of the thermal advection by these two wind components and their sum at the 950 hPa level on day 6.5 in the Control Run. The wind fields are overlaid in the plots. Panel (a) and (b) show that both components give rise to cold advection in the cold front region and hence would reinforce one another. In the warm front region, while the geostrophic wind gives rise to *warm advection*, the ageostrophic wind gives rise to a somewhat weaker *cold advection*. The partial compensation results in a net warm advection (panel (c)). These features would imply a large local rate of change of temperature in both cold and warm front regions of opposite signs. The result in this general model corroborates the finding obtained in the counterpart analysis with a dry 3D SG model discussed in Chapter 5. We will further show the dynamical differences underlying the formation of the two frontal regions.

7.4.5 Rate of Surface Frontal Development

As discussed in Section 1.3, the rate of surface frontogenesis can be appropriately measured as $F_2 = \frac{D}{Dt}|\nabla_2\theta|$. We will examine this measure in conjunction with the decomposition of the wind field into its geostrophic and ageostrophic components. The symbolic form of this metric for making computation directly with the data on pressure levels is (1.7).

Horizontal Distribution of F_2 Over Surface Fronts

We reproduce Figure 7.6b here as Figure 7.10a for the convenience of making comparison. It shows the distribution of F_2 at the 925 hPa level on day 6.5 when the fronts are intensifying most rapidly with diabatic heating in the Control Run. The three parts of F_2 are shown in panels (b), (c) and (d) of Figure 7.10. The red contours in Figure 7.10b depict close collation of the three-hour accumulated precipitation and the surface fronts. The former has a maximum value of 22 mm. The strong diabatic heating takes place at and near the surface fronts themselves, with a large spatial gradient. The direct impact of diabatic heating F_{2H} and the effect of the total flow field (the sum of F_{2G} and F_{2A}) are comparably important in surface frontogenesis. F_{2A} is considerably greater than F_{2G} in the cold front region, suggesting that the ageostrophic velocity component plays a greater role in cold frontogenesis. The contributions from the geostrophic velocity component F_{2G} and those from the ageostrophic component F_{2A} are comparable in the warm front region, albeit the latter is still slightly stronger. The structure is rather complex since it consists of both broad bands as well as scattering small patches. There are also some additional scattered

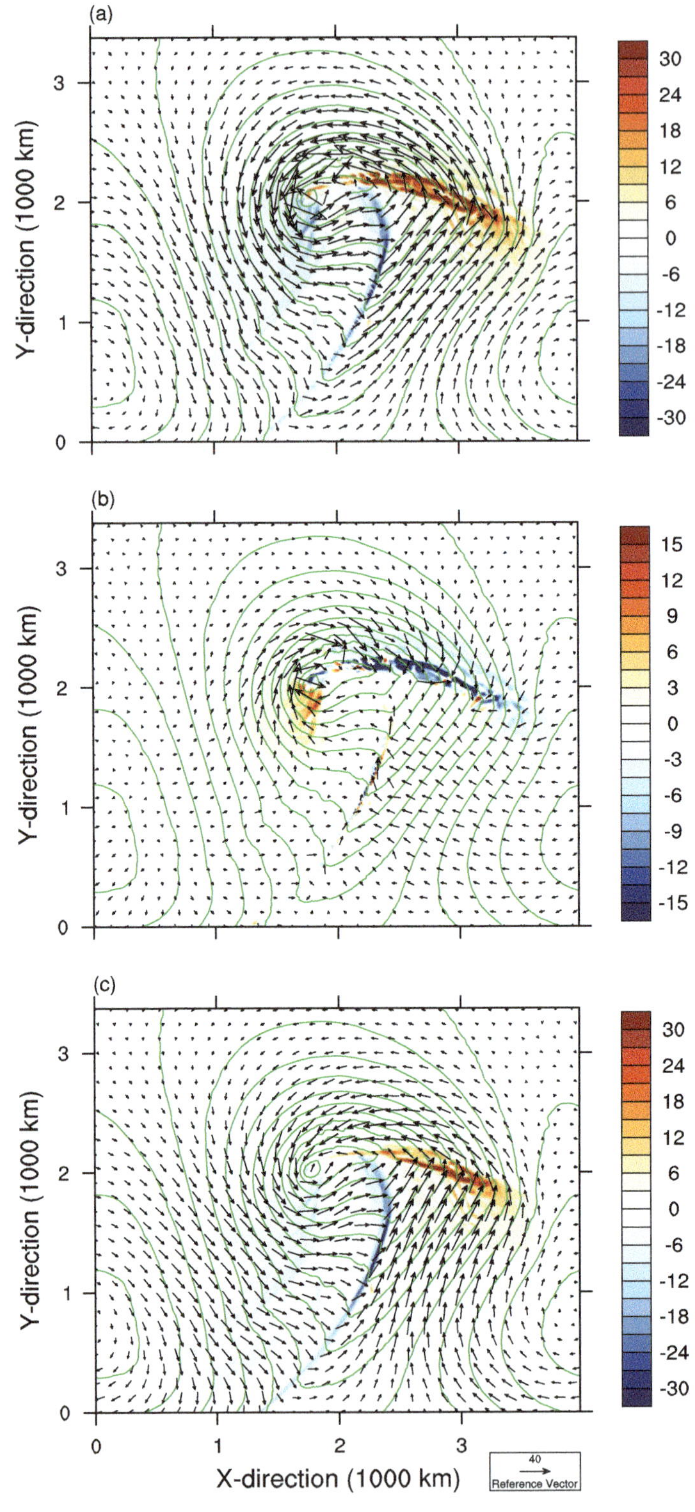

Figure 7.9. Distributions of the velocity (vectors), surface pressure (contours, min. = 974 hPa, CI = 4 hPa) and its thermal advection (shading) in 1×10^{-4} K s^{-1} at 950 mb on day 6.5 in the Control Run by (a) geostrophic velocity component, (b) ageostrophic velocity component, and (c) their sum (taken from MLD2).

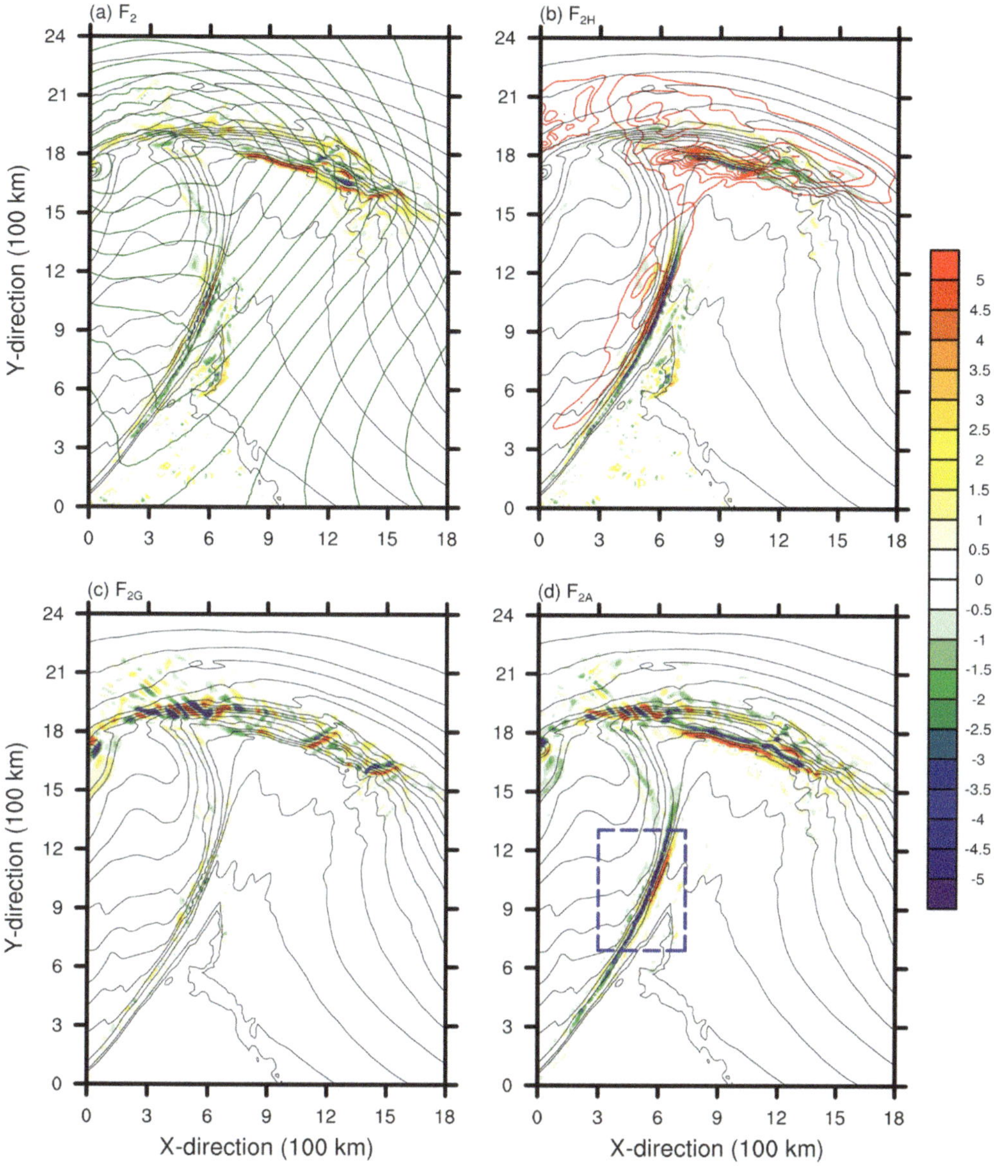

Figure 7.10. Distributions of (a) F_2 (shading) in 10^{-8} K m^{-1} s^{-1} with surface pressure (green contours, CI = 6 hPa) and potential temperature θ with 3-hour (09:00 to 12:00 of day 6) accumulated precipitation rate in millimeters (red contours; CI = 2 mm, max. 22 mm), (c) F_{2G} with θ and (d) F_{2A} with θ at 925 hPa on day 6.5 in the Control Run. The blue box indicates a small segment of the cold front to be further examined in Figure 7.11 (taken from MLD2).

patches of values of both signs in F_2 in the warm sector east of the cold front. They are associated with the diabatic heating processes as indicated in panel (b).

The four panels of Figure 7.11 give a close-up look at the fields of θ, F_2, F_{2H}, F_{2G}, and F_{2A} in a small segment of the cold front indicated by the blue box in

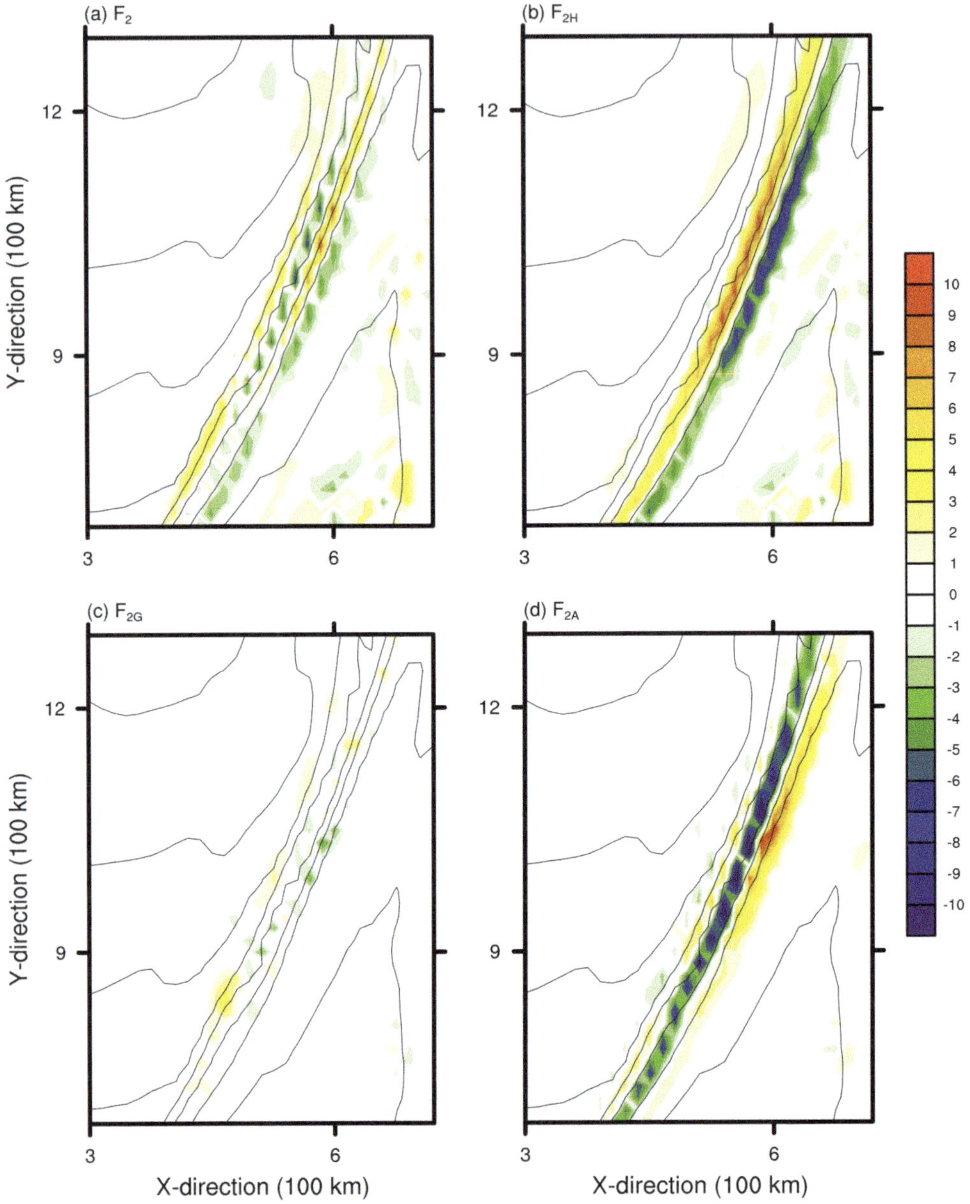

Figure 7.11. Distribution of potential temperature with (a) F_2 across the cold front at 925 hPa on day 6.5 in 10^{-8} K m s^{-1}, (b) F_{2H}, (c) F_{2G}, and (d) F_{2A} zoomed in an enlarged small segment of the cold front indicated by the blue box in Figure 7.10d, depicting the superfine structure (taken from MLD2).

Figure 7.10d. They reveal the superfine structure across the cold front. The values of F_{2G} are the smallest even though the geostrophic flow component is the primary part of the circulation in a baroclinic unstable wave. The differences in the bands in F_{2A} and F_{2H} are particularly noteworthy. The pronounced red band (positive values)

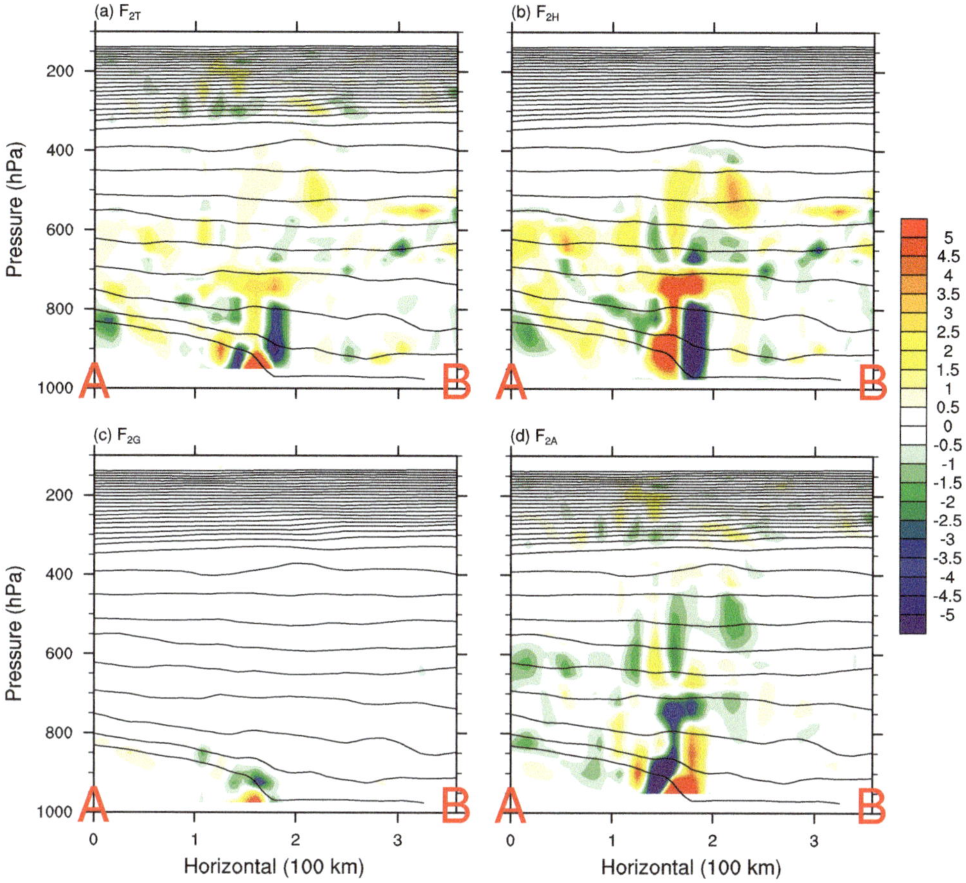

Figure 7.12. Vertical distributions of F_2 (shading) and potential temperature (contours) passing through AB over the cold front on day 6.5 in Control Run: (a) F_2, (b) F_{2H}, (c) F_{2G}, and (d) F_{2A}.

in F_{2H} is located to the west of the blue band (negative values), suggesting that diabatic heating tends to move the front westward. In contrast, the pronounced red band in F_{2A} is located to the east of the blue band, suggesting that the ageostrophic flow component tends to move the front eastward. There is considerable cancellation between F_{2H} and F_{2A} in the cold front region, making the bands in F_2. even narrower. The net effect contributes to the narrowing of the cold front. The effect of heating is greatly canceled by the effect of ageostrophic flow at the warm front, leading to a relatively weak and broad warm frontal zone.

Vertical Distribution of F_2 Over the SFs

To ascertain the factors responsible for the vertical structure of the SFs, we next examine the vertical distribution of F_2 in a cross-section passing through the two lines AB and CD indicated in Figure 7.7a. Figure 7.12 shows the distributions of

F_2 and its three constituent parts of F_2 together with the potential temperature over the cold front along AB on day 6.5. It reveals that the cold frontal layer extends to about 850 hPa behind its leading edge. The nearly uniform potential temperature is an indication of considerable vertical mixing within such layer, which is capped by highly stratified air at its top. The range of values of F_2 is $\pm 10 \times 10^{-8}$ K m^{-1} s^{-1}. The positive values of F_2 at the leading edge of the cold front seen in panel (a) are mostly attributable to the F_{2A} term, as evident in panel (d). There are also values of alternating signs along the top of this relatively thin cold front layer. It is noteworthy that the values of F_{2G} are considerably smaller, even though the geostrophic flow component is the primary part of the circulation in a baroclinic unstable wave. We conclude that the ageostrophic velocity component plays a more significant role as far as frontogenesis is concerned. The diabatic heating term also has significant values in the lower half of the troposphere. There is noticeable cancelation between the F_{2H} and F_{2A} terms, leading to smaller net values in F_2.

In a cross-section over the warm front region along the line CD, we see in Figure 7.13 comparable contributions from the diabatic heating and the ageostrophic velocity in much of the lower half of the troposphere. There seems to be partial cancelation between the F_{2H} and F_{2A} terms, leaving rather small net values of F_2 at most levels. Again the geostrophic velocity component is found to play a minor role.

7.4.6 Characteristics of the Constituent Parts of F_{2G} and F_{2A}

To further delineate the processes underlying F_{2G} and F_{2A} in the dynamics of surface frontogenesis, we examine each of them as functions of their deformation properties. We may rewrite F_{2G} as

$$F_{2G} \equiv -\frac{1}{|\nabla_2\theta|}\nabla_2\theta\cdot\boldsymbol{R} = -\frac{1}{|\nabla_2\theta|}\frac{1}{2}(\theta_x^2-\theta_y^2)D_1 - \frac{1}{|\nabla_2\theta|}\theta_x\theta_y D_2$$
$$\equiv F_{2G}(D_1) + F_{2G}(D_2), \tag{7.1}$$

where $D_1 = \hat{u}_x - \hat{v}_y$ and $D_2 = \hat{u}_y + \hat{v}_x$ are the stretching deformation and shearing deformation of the geostrophic velocity $(\hat{u},\ \hat{v})$. This equation tells us that F_{2G} mostly stems from the shearing deformation of the geostrophic flow component at the locations where the isotherms are roughly in the SW–NE direction ($|\theta_x| \sim |\theta_y|$), as in the cold front. On the other hand, where the isotherms are either close to the zonal direction ($\theta_x \sim 0$) or meridional direction ($\theta_y \sim 0$), F_{2G} would be mostly attributable to the stretching deformation as in the warm front. Figures 7.14a,b show respectively the distribution of $F_{2G}(D_1)$ and $F_{2G}(D_2)$ at 925 hPa on day 6.5 in the Control Run. We find small negative values for $F_{2G}(D_1)$ and somewhat larger positive values for $F_{2G}(D_2)$ in the cold front region, suggesting that shearing deformation of the geostrophic velocity component mostly contributes to the cold front formation. The values of $F_{2G}(D_1)$ and $F_{2G}(D_2)$ in the warm front region have smaller scales.

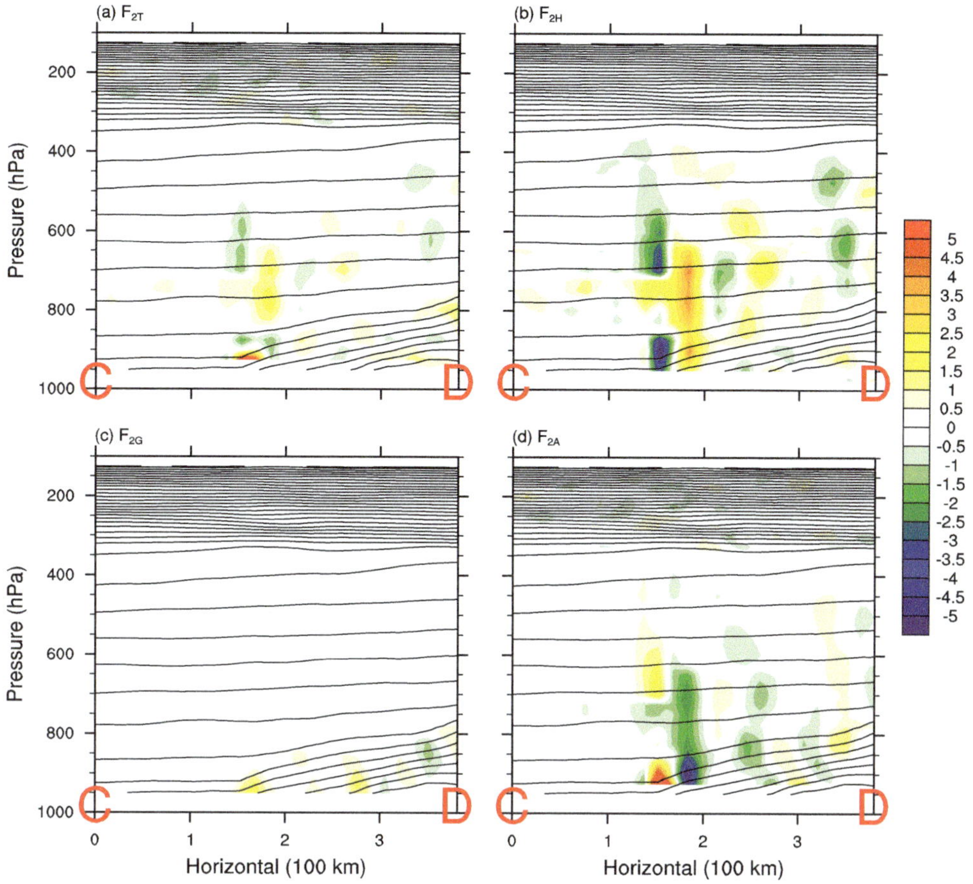

Figure 7.13. Vertical distributions of F_2 (shading) and potential temperature (contours) passing through CD over the cold front on day 6.5 in the Control Run: (a) F_2, (b) F_{2H}, (c) F_{2G} and (d) F_{2A}.

The deformation properties of a 3D ageostrophic flow may be defined on three orthogonal planes. The stretching and shearing deformations of the 3D ageostrophic wind component on the (x–y) plane are $I_1 = \tilde{u}_x - \tilde{v}_y$ and $I_2 = \tilde{u}_y + \tilde{v}_x$; those on the ($x$–$\tilde{z}$) plane are $J_1 = \tilde{u}_x - \tilde{w}_{\tilde{z}}$ and $J_2 = \tilde{u}_{\tilde{z}} + \tilde{w}_x$; those on the ($y$–$\tilde{z}$) plane are $K_1 = \tilde{v}_y - \tilde{w}_{\tilde{z}}$ and $K_2 = \tilde{v}_{\tilde{z}} + \tilde{w}_y$. We may rewrite the terms in F_{2A} into two subgroups as follows

$$F_{2A} \equiv -\frac{1}{|\nabla_2\theta|}\nabla_2\theta \cdot \boldsymbol{B} = F_{2A}(I) + F_{2A}(II), \tag{7.2}$$

$$F_{2A}(I) = -\frac{1}{|\nabla_2\theta|}(\theta_x^2 J_1) - \frac{1}{|\nabla_2\theta|}(\theta_y^2 K_1) - \frac{1}{|\nabla_2\theta|}\theta_x\theta_y I_2$$

$$F_{2A}(II) = -\frac{1}{|\nabla_2\theta|}\left[(\theta_x^2 + \theta_y^2)\tilde{w}_{\tilde{z}} + \theta_{\tilde{z}}\theta_x\tilde{w}_x + \theta_{\tilde{z}}\theta_y\tilde{w}_y\right].$$

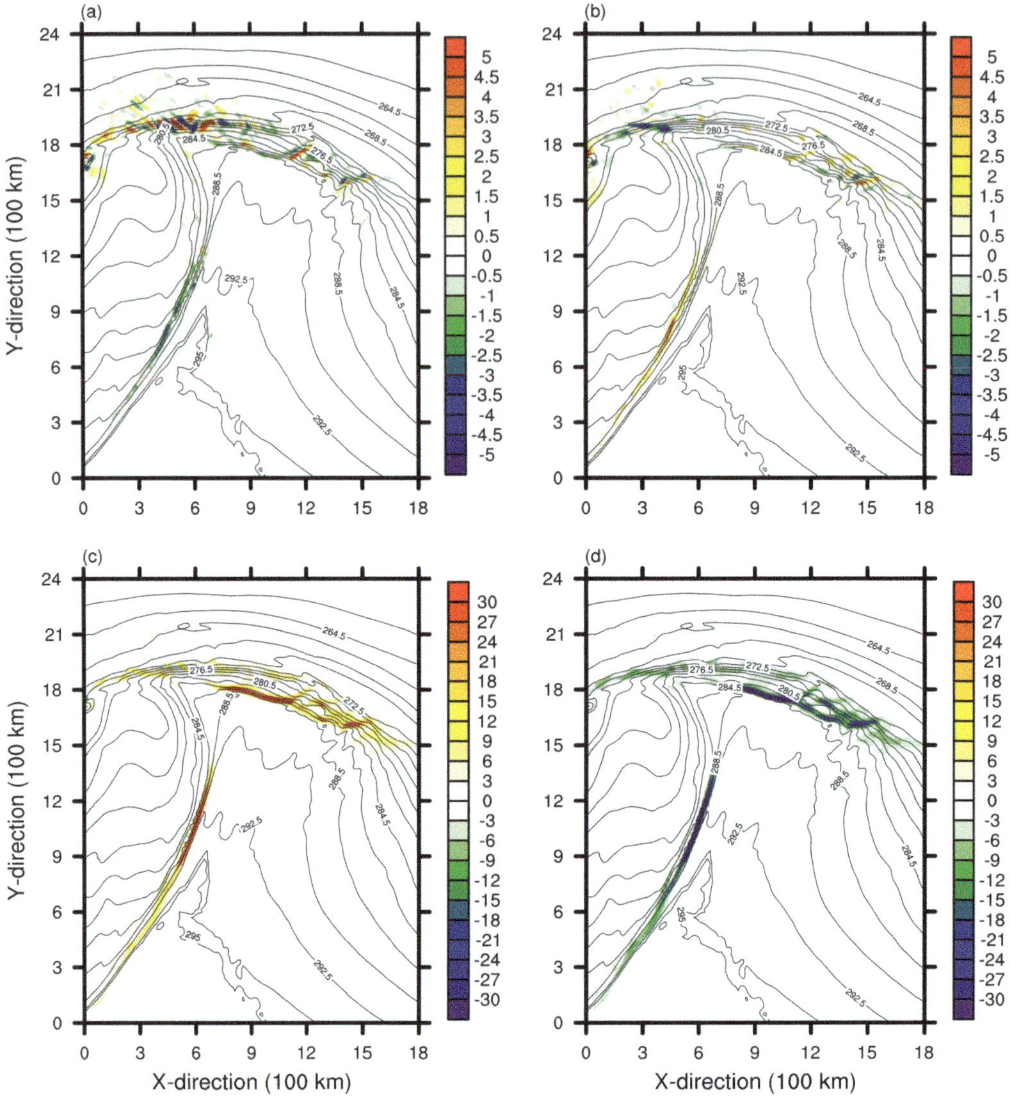

Figure 7.14. Distribution of (a) $F_{2G}(D_1)$, (b) $F_{2G}(D_2)$, (c) $F_{2A}(I)$ and (d) $F_{2A}(II)$ (shading) in 10^{-8} K m^{-1} s^{-1} at 925 hPa on day 6.5 in the Control Run. Superimposed on each panel is also potential temperature (contours; CI = 2 K) (taken from MLD2).

The term $F_{2A}(I)$ is a measure of the total influence of the deformation property of the 3D ageostrophic wind vector. The term $F_{2A}(II)$ may be interpreted as a form of 3D tilting process. The partition of the terms in (7.2) is not unique, and is only meant as broad guidance for making interpretations. For instance, $\tilde{w}$ is not exclusively present in the term $F_{2A}(II)$, for it also appears in K_1.

Figures 7.14c,d show respectively the horizontal distributions of $F_{2A}(I)$ and $F_{2A}(II)$ at 925 hPa on day 6.5 in the Control Run. The values are much larger than those of $F_{2G}(D_1)$ and $F_{2G}(D_2)$ since the range of the color bar used in these two panels is six times larger. Figure 7.14c reveals that the term $F_{2A}(I)$ involving the three parts of deformation has *positive* values in both the cold front and warm front. In contrast, the term $F_{2A}(II)$ shown in Figure 7.14d has large *negative* values in both cold and warm fronts. In other words, the 3D tilting process is frontolytic, whereas the 3D deformation process is frontogenetic as far as the 3D ageostrophic wind component is concerned. The high degree of cancelation of these two effects gives rise to the fine bands of alternating positive and negative values seen in Figure 7.11d earlier. The plots of the individual parts of $F_{2A}(I)$ (not shown for brevity) reveal that the term proportional to $J_1 = \tilde{u}_x - \tilde{w}_z$ has the largest positive values along the cold front, whereas the term proportional to $K_1 = \tilde{v}_y - \tilde{w}_z$ has the largest positive values in the warm front region. The term involving I_2 has relatively small values, although they are positive. We have also computed the vertical distributions of these constituent parts of F_{2G} and F_{2A}. It is found that the term $F_{2A}(I)$ makes the largest positive contribution in the warm front layer, but is largely counteracted by the term $F_{2A}(II)$. The detailed results are not presented for the sake of brevity.

We have also computed the vertical distributions of these constituent parts of F_{2G} and F_{2A}. It is found that $F_{2A}(I)$ makes the largest positive contribution in the warm front layer, but is largely counteracted by $F_{2A}(II)$. The detailed results are not presented for the sake of brevity.

7.5 Upper-Level Frontogenesis

The ULF in the Dry Run is weaker but similar to that in the Control Run. We will first present the results concerning the properties of ULF in the Dry Run. We begin by going over the rationale and derivation of a frontogenetical function for ULF to be referred to as F_3 in Section 7.5.1. In Section 7.5.2, we present the structural characteristics of the ULF. In Section 7.5.3, we diagnose the rate of development of the corresponding ULF in terms of F_3. In Section 7.5.4, we examine the transverse circulation associated with that ULF. In Section 7.5.5, we make a closer examination of the constituent parts of F_3. In Section 7.5.6, we briefly present evidence for the existence of internal gravity waves in the model simulation. Finally, in Section 7.5.7, we discuss the impacts of diabatic heating on the ULF on the basis of the simulation in the Control Run.

7.5.1 Frontogenetical Function for the ULF

The ULF is characterized by locally large magnitude in the 3D gradient of potential temperature, $|\nabla\theta|$, as seen in Figures 1.1 and 1.2. The corresponding frontogenetical function would be $F_3 \equiv \frac{D}{Dt}|\nabla\theta|$. There is a brief discussion of F_3 in the textbook by Bluestein (1993: section 2.3.2), but it was applied for the first time in MLD1. One might question whether $\frac{D}{Dt}|\nabla\theta|$ is a suitable measure as a frontogenetical function for ULF on the grounds that a tropopause, which is characterized by large vertical derivatives of potential temperature alone, is generally not regarded as a front. It is relevant to note that a background tropopause and a ULF differ greatly in terms of their spatial and temporal scales, arising from entirely different physical processes. Two primary factors set the stage for the existence of tropopause in middle and high latitudes: solar radiative heating at the Earth's surface and absorption of ultraviolet radiation by ozone in the stratosphere. The terrestrial radiative cooling within the troposphere and the vertical heat transports by numerous cycles of small-scale convection as well as baroclinic waves control the mean lapse rate and thereby the height of the tropopause in the extratropics. Such resulting tropopause is essentially horizontal and has a planetary scale. In the tropics, the tropopause is distinctly higher because it largely reflects the vertical extent of the Hadley cell. The tropical mean tropopause then has a planetary scale in the zonal direction. The Hadley circulation stems from the meridional differential solar heating, long-wave radiative cooling in the subtropical troposphere and the self-induced condensational heating in the equatorial latitudes. The meridional extent of the Hadley cell in turn is dictated by the characteristics of its dynamic stability. A byproduct of the Hadley circulation is a baroclinic jet just beyond the poleward side of the subtropics due to quasi-conservation of angular momentum of the poleward moving air parcels. It follows that the mean tropopause would slope steeply poleward over the subtropics and have a seasonal time scale. These factors can be implicitly incorporated through a reference state of our model to be used in this chapter. In contrast, ULF is associated with local and transient *tropopause folding* as an element of an intensifying baroclinic wave. It follows that *ULF has synoptic and subsynoptic scales in the departure field* from our reference state with a significant 3D gradient of potential temperature. The latter is easily distinguishable from the largely seasonal and planetary scales in the fluctuation of the extratropical tropopause. Therefore it would be reasonable to quantify ULF in terms of the 3D variation of the potential temperature, $|\nabla\theta|$. We refer to the corresponding metric $F_3 \equiv \frac{D}{Dt}|\nabla\theta|$ as the "frontogenetical function in 3D space" for short. *It should be noted that F_2 has been traditionally used as the frontogenetical function for ULF. Readers are referred to the many references cited by* Buzzi (2017) *in his Comment on MLD1. His remarks concerning this matter and the reply to his comment by MLD is a dialogue of pedagogical interest (*Mak et al., 2017).

In working with data available at different pressure levels, we wish to use it directly without interpolating the data. We would need to perform the diagnoses in a coordinate system that has the same unit in all three directions. For this purpose, we introduce a modified p-coordinate denoted by $\tilde{z}$ in units of m. In such $(x, y, \tilde{z})$ coordinates, we would have $\nabla = \left(\frac{\partial}{\partial x}, \frac{\partial}{\partial y}, \frac{\partial}{\partial \tilde{z}}\right)$ in units of m^{-1} in each component and $\boldsymbol{V} = (u, v, \tilde{w})$ with $\tilde{w} = \frac{D\tilde{z}}{Dt}$ in units of $\mathrm{m\,s}^{-1}$. An appropriate choice of $\tilde{z}$ can be

$$\tilde{z} = z_{top}\left(1 - \frac{p}{g\rho_o}\right), \tag{7.3}$$

$$\tilde{w} = -\left(\frac{\omega}{g\rho_o} + \frac{p}{g}\frac{D}{Dt}\left(\frac{1}{\rho_o}\right)\right), \tag{7.4}$$

where ρ_o is density which is a prescribed function of p compatible with the observed atmospheric mean density. Thus, $\tilde{z}$ is a height-like coordinate. This coordinate is compatible with a mean atmosphere that is approximately in hydrostatic balance. One may introduce alternative coordinates instead that would serve the same purpose, e.g., the so-called pseudo-height coordinate (Hoskins and Bretherton, 1972).

It is straightforward to work out an explicit expression for $\frac{D}{Dt}|\nabla\theta|$ in $(x, y, \tilde{z})$ coordinates. With the use of the thermodynamic equation, $\frac{D\theta}{Dt} = \frac{\theta H}{c_p T}$, where H is the net heating rate in units of $\mathrm{K\,s}^{-1}$, it can be readily shown that the explicit form of F_3 in units of $\mathrm{K\,m}^{-1}\,\mathrm{s}^{-1}$ is:

$$\begin{aligned}\frac{D}{Dt}|\nabla\theta| = \frac{1}{|\nabla\theta|}\Bigg\{&\theta_x\left[\frac{1}{c_p}\left(\frac{p_o}{p}\right)^{\kappa}\frac{\partial H}{\partial x} - u_x\theta_x - v_x\theta_y - \tilde{w}_x\theta_{\tilde{z}}\right]\\ &+\theta_y\left[\frac{1}{c_p}\left(\frac{p_o}{p}\right)^{\kappa}\frac{\partial H}{\partial y} - u_y\theta_x - v_y\theta_y - \tilde{w}_y\theta_{\tilde{z}}\right]\\ &+\theta_{\tilde{z}}\left[\frac{1}{c_p}\frac{\partial}{\partial \tilde{z}}\left(\left(\frac{p_o}{p}\right)^{\kappa}H\right) - u_{\tilde{z}}\theta_x - v_{\tilde{z}}\theta_y - \tilde{w}_{\tilde{z}}\theta_{\tilde{z}}\right]\Bigg\}.\end{aligned} \tag{7.5}$$

It should be added that H not only includes the contributions from all forms of diabatic heating processes, but also the conversion from kinetic to thermal energy due to turbulent mixing. If we decompose the horizontal velocity as $u = \hat{u} + \tilde{u}$, $v = \hat{v} + \tilde{v}$, with $(\hat{u}, \hat{v})$ being the geostrophic component and $(\tilde{u}, \tilde{v})$ the ageostrophic component, the 12 terms in (7.5) would become 16 terms. Let us introduce a 2D vector function of the geostrophic velocity $(\hat{u}, \hat{v})$ and θ:

$$\boldsymbol{S} = (\nabla\hat{u}\cdot\nabla\theta, \quad \nabla\hat{v}\cdot\nabla\theta), \tag{7.6}$$

and a 3D vector function of the 3D ageostrophic velocity $(\tilde{u}, \tilde{v}, \tilde{w})$ and θ:

$$\boldsymbol{A} = (\nabla\tilde{u}\cdot\nabla\theta, \quad \nabla\tilde{v}\cdot\nabla\theta, \quad \nabla\tilde{w}\cdot\nabla\theta). \tag{7.7}$$

Then, (7.5) can be rewritten concisely as

$$\frac{D}{Dt}|\nabla\theta| = \frac{1}{c_p|\nabla\theta|}\nabla\theta\cdot\nabla\left(\left(\frac{p_o}{p}\right)^{\kappa}H\right) - \frac{1}{|\nabla\theta|}\nabla_2\theta\cdot S - \frac{1}{|\nabla\theta|}\nabla\theta\cdot A. \tag{7.8}$$

The ageostrophic velocity component is expected to play a different but equally important role in spite of being generally much weaker than the geostrophic velocity.

It should be emphasized that no physical or mathematical approximations have been invoked in getting (7.5) and hence (7.8), although one would unavoidably have to introduce some forms of approximation in making evaluation of H. Thus, there is no restriction to the validity of (7.8) in principle. For convenience, let us refer to (7.8) symbolically as follows:

$$F_{3T} = F_{3H} + F_{3G} + F_{3A}, \tag{7.9}$$

where $F_{3H} = \frac{1}{|\nabla\theta|}\nabla\theta\cdot\nabla\left(\frac{H}{c_p}\left(\frac{p_o}{p}\right)^{\kappa}\right)$, $F_{3G} = -\frac{1}{|\nabla\theta|}\nabla_2\theta\cdot\boldsymbol{S}$, and $F_{3A} = -\frac{1}{|\nabla\theta|}\nabla\theta\cdot\boldsymbol{A}$. F_{3H} is the effect of net diabatic heating on the rate of frontogenesis when the latter is measured in terms of $\frac{D}{Dt}|\nabla\theta|$. F_{3G} and F_{3A} are measures of the dynamical effect of the geostrophic component and ageostrophic component of the flow on frontogenesis respectively. $F_2 \equiv \frac{D}{Dt}|\nabla_2\theta|$ is a special version of $F_3 \equiv \frac{D}{Dt}|\nabla\theta|$. In Section 7.6.3 we will use (7.9) and other related formulas to diagnose the dynamical nature of upper-level frontogenesis.

7.5.2 Structural Characteristics of the ULF in the Dry Run

Upper-level frontogenesis in our Dry Run can be most succinctly diagnosed from the perspective of PV for it is largely a conservative material property in the absence of diabatic heating. The generic definition of PV is $q = \frac{1}{\rho}(f\boldsymbol{k} + \nabla\times\vec{V})\cdot\nabla\theta$, where $\boldsymbol{k}$ is the vertical unit vector and other notations have standard meanings. Using the model data on pressure levels, we compute q with the following formula written on $(x, y, \tilde{z})$ coordinates elaborated in Section 7.2:

$$q = \frac{1}{\rho_o}\{(\tilde{w}_y - v_{\tilde{z}})\theta_x + (u_{\tilde{z}} - \tilde{w}_x)\theta_y + (f + v_x - u_y)\theta_{\tilde{z}}\}, \tag{7.10}$$

in units of PVU (1 PVU $= 10^{-6}$ m^2 s^{-1} K(kg)$^{-1}$). The immediate task is to establish the existence of model ULF and delineate its structure. Since it is a 3D disturbance, a concise and effective way of doing so is to make judiciously chosen horizontal and vertical cross-sections of the PV field. The location of $x = 0$ in this section concerning the Dry Run has been shifted westward by 190 grid points for presentation purposes. We show the height field and PV field of the flow at two interior levels, 500 hPa and 700 hPa, on day 6.5. It is seen in Figure 7.15a that this synoptic mid-tropospheric wave is quite intense, with the lowest contour depicting a cutoff low. The range of the height field on 500 hPa is 5100–5800 m. Particularly informative is the PV field,

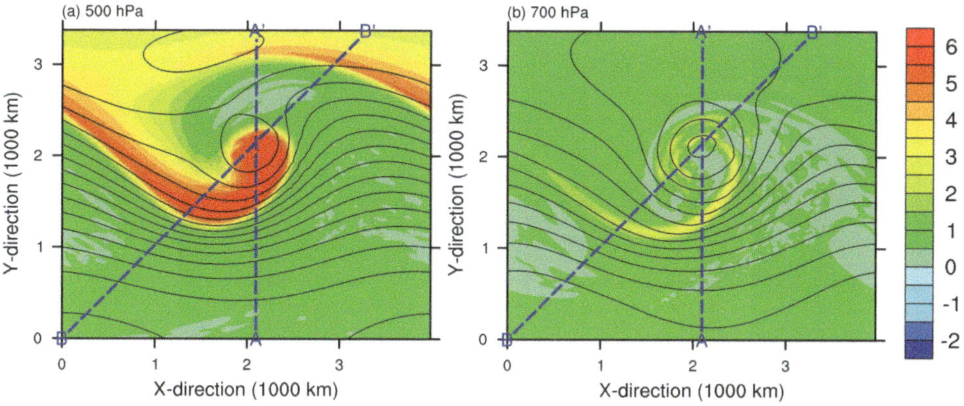

Figure 7.15. Distributions of PV (shading) in PVU and geopotential height (contours) in meters at (a) 500 mb level and (b) 700 mb on day 6.5 of the Dry Run (taken from MLD1).

which ranges from 0.4 to 6 PVU. There is a narrow tongue of large values of PV across the mid-section of the trough. Notice that the brightest red color (and hence largest PV value) appears at the southern edge of this penetrating tongue. These large PV values could not simply have arisen from horizontal advection of PV. The air parcels with greater than 1.5 PVU are of stratospheric origin and must have sunk through the tropopause to this level. A corresponding tongue of relatively large PV values is also evident at 700 hPa (Figure 7.15b). The maximum value here is about 3.5 PVU, implying that the air in this strip with this PV value is also of stratospheric origin. Taken together, these two panels reveal that an integral part of the wave development is to bring down some air parcels from certain stratospheric levels to middle and even lower tropospheric levels in a limited strip of area across the upstream of the baroclinic wave trough. It is also noteworthy that the wave has already intensified to the maximum extent by this time since it has virtually no vertical tilt left at this point.

The distribution of the magnitude of the *spatial derivative* of potential vorticity on a pressure surface, $|\nabla_2 q|$, would provide a basis to estimate the width of the ULF at that level. Figure 7.16 shows a ULF with a curvilinear structure at the 500 hPa level. The part near the cutoff low extends down to 700 hPa. The maximum value is found in a segment to the south of the cutoff low at about $50 \times 10^{-12}(\text{m}^2\,\text{s}^{-1}\,\text{kg}^{-1}\,\text{K})\text{m}^{-1} = 5\,\text{PVU}/(100\,\text{km})$. The width of the ULF therefore varies somewhat along it. The narrowest width of the model ULF at that level is therefore estimated to be about 100 km.

Next, we show a plot of the vertical distributions of PV overlaid on the potential temperature field in a cross-section passing through the line AA′ as well as another plot through the line BB′. Those lines are indicated in Figure 7.15. The line AA′ is oriented in a north–south direction, passing through the cutoff low. The line BB′ is

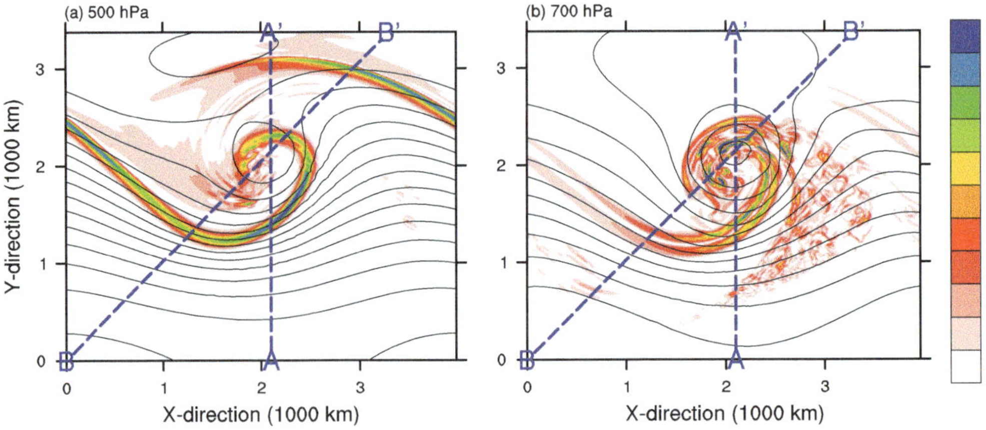

Figure 7.16. Distributions of $|\nabla_2 q|$ on (a) 500 hPa and (b) 700 hPa surfaces in units of 10^{-12} m s^{-1} kg^{-1} K on day 6.5 in the Dry Run.

in the SW–NE direction cutting through the disturbance center. Such distributions are shown in Figure 7.17. Figure 7.17a clearly reveals a ULF that manifests as a tongue of high PV values extending downward from the tropopause at a location about 1200 km north of point A. It is collocated with the steep sloping contours of potential temperature there. There are similar characteristics in the BB′ cross-section in Figure 7.17b. This ULF extends down to about 700 hPa on the AA′ cross-section and to about 600 hPa on the BB′ cross-section. These two panels by themselves might misleadingly suggest that there exist three ULFs, each originating from three locations of the perturbed tropopause. But we have seen in Figure 7.16 that they are in fact just three parts of a single ULF. It is noteworthy that there is also a small patch of high PV values at the lower tropospheric levels. That arises from the boundary layer turbulent mixing, which is treated in the model as a form of diabatic heating.

There is a large change in density across each point on the ULF since pressure must vary continuously at any point on a ULF across which there is a jump in density and potential temperature. Let us denote the pressure, density and potential temperature of the air of tropospheric origin at a point immediately next to the ULF by p_*, ρ_* and θ_* respectively. Then these properties of the air of stratospheric origin at a point next to the ULF are p_*, $(\rho_* + \Delta\rho)$ and $(\theta_* + \Delta\theta)$. It follows from the equation of state for dry air we would have $\Delta\rho = -\frac{\rho_*}{\theta_*}\Delta\theta$ where $\rho_* = (p_*^{\gamma} p_{oo}^{\kappa})/(R\theta_*)$, $\kappa = R/c_p$, and $\gamma = c_v/c_p$. Hence, there is a large decrease in density proportional to the large increase in potential temperature across a ULF. In other words, the ULF and SF are both transition zones between two air masses of different density. While the SF separates air masses of high and low latitude origins at the Earth's surface, the ULF separates air masses of stratospheric and tropospheric origins aloft.

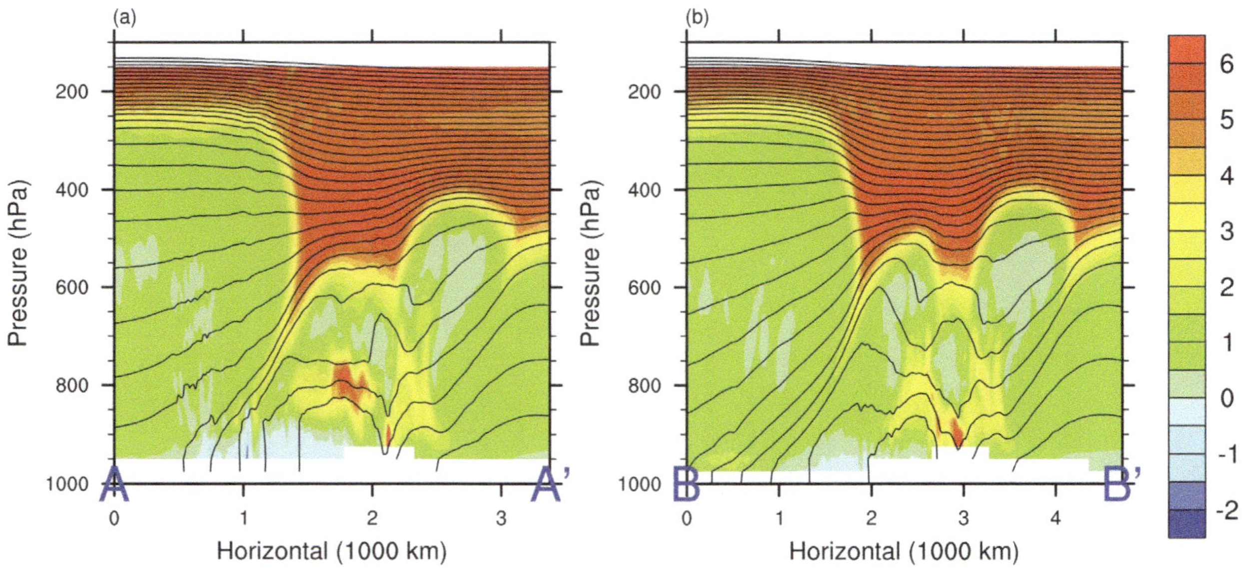

Figure 7.17. Distributions of PV (shading in PVU) and potential temperature (contours) in K on the vertical cross-section passing (a) through AA′ and (b) through BB′ on day 6.5 of the Dry Run (taken from MLD1).

To confirm that there is systematic descending motion in the region of this ULF, we plot the distribution of vertical velocity on the cross-sections along AA′ and BB′. The results are shown in Figure 7.18. The structure of the vertical velocity field in the BB′ cross-section is particularly distinct. This is synoptic-scale descent of the baroclinic wave reaching a maximum value of about 10 cm s^{-1} at about 550 hPa. In passing, we also note that there is synoptic-scale ascent on the eastern side of the trough, especially in the wrap-around region as expected in a baroclinic wave.

7.5.3 Rates of ULF Development

The signature feature of a ULF associated with tropopause folding is the large value of the 3D gradient of potential temperature. A logical metric for measuring its rate of development, $F_3 \equiv \frac{D}{Dt}|\nabla\theta|$, was formulated in Section 7.5.1. The distributions of F_3 and its parts on the north–south-oriented vertical cross-section passing through AA′ on day 6.5 in the Dry Run are shown in Figure 7.19. Note that we use different color bars in the plotting of F_{3T}, F_{3H}, F_{3G} and F_{3A} in order to bring out the detailed features in each. F_{3A} is so much larger than F_{3H} and F_{3G} that it is virtually equal to F_{3T}. We see that the large positive values of F_{3T} collocate well with the ULF on this cross-section. The maximum value is about 1000 units of 10^{-8} K s^{-1} m^{-1}. The ULF is located in a lower tropospheric layer near the 700 hPa level and about 1200 km north of point A. The large negative values are found in a location of decreasing gradient of potential temperature, i.e., a site of frontolysis. Figure 7.19 reveals that F_3 is almost entirely attributable to the dynamic influence of the ageostrophic velocity. The much smaller values in the heating term of this Dry Run solely stem from the turbulent heat flux represented in the PBL scheme of the model. The direct influence of the geostrophic flow component is also small, although its indirect influence through the synoptic-scale transverse circulation of the intensifying baroclinic wave is significant. There is a hint of the presence of vertically propagating gravity waves at the stratospheric levels originated from the surface cold front. A closer diagnosis of this will be discussed shortly.

The counterpart result of F_3 and potential temperature on the vertical cross-section in a general orientation passing through BB′ is shown in Figure 7.20. It shows an even more definitive correlation between the large positive values of F_{3T} with the location of the ULF and between the large negative values with the location of decreasing gradient of potential temperature. The significant positive values appear in the whole tropospheric layer from 900 hPa to the tropopause. It is also entirely attributable to the impact of the ageostrophic velocity component. They are located near the center of the disturbance (at ~2000 km from point A and ~3000 km from point B). They, too, are associated with the impact of the ageostrophic flow component.

Figure 7.21 shows the *horizontal distribution* of F_3 and its three parts, together with the potential temperature on 700 hPa. The horizontal gradient of potential

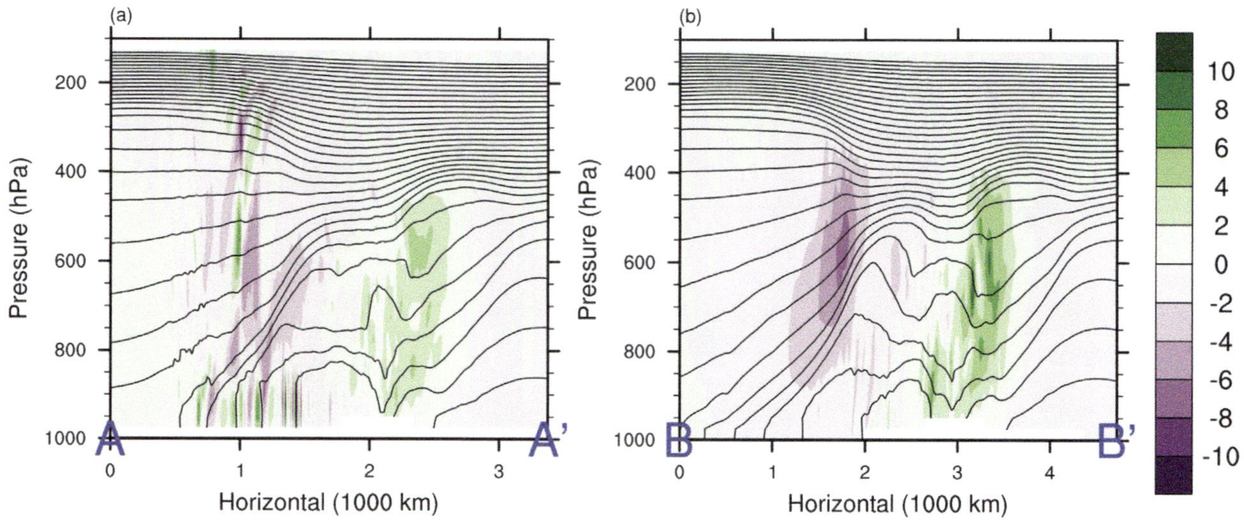

Figure 7.18. Distributions of vertical velocity (shading in cm s^{-1}) and potential temperature (contours) in K on the vertical cross-section passing (a) through AA′ and (b) through BB′ on day 6.5 of the Dry Run (taken from MLD1).

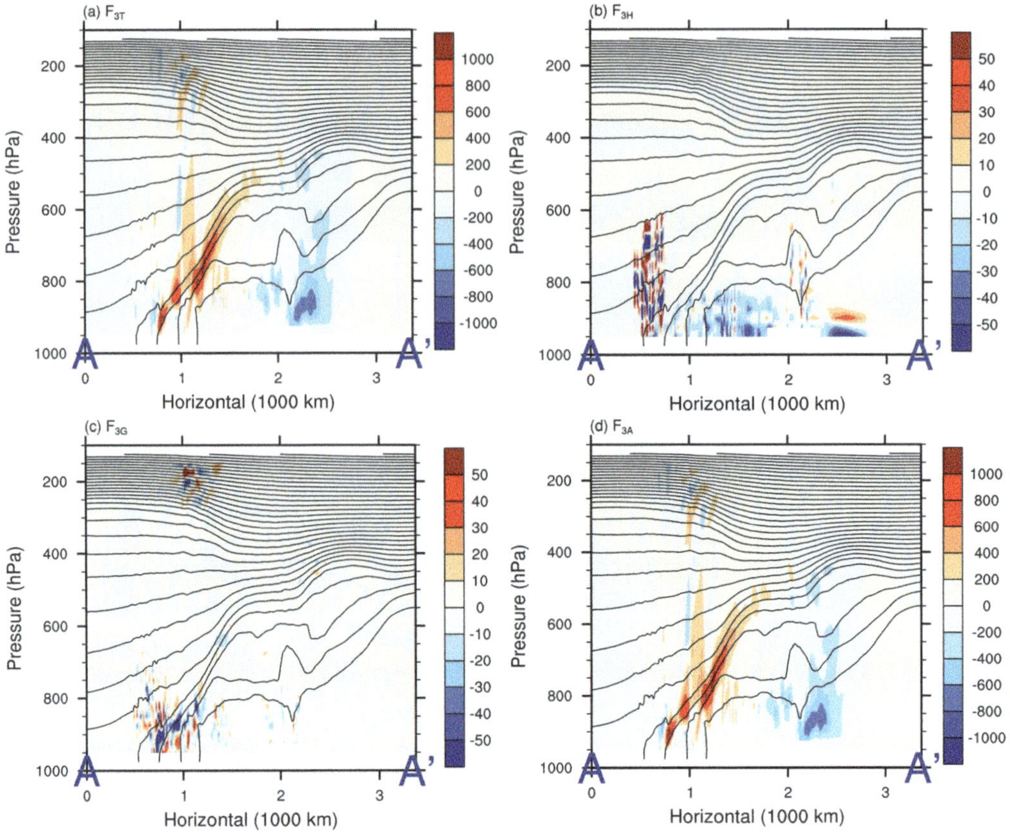

Figure 7.19. Distributions of F_{3T} and its three components F_{3H}, F_{3G}, F_{3A} (shade), in units of 10^{-8} K s^{-1} m^{-1} and potential temperature (contours) in K on the cross-section AA′ on day 6.5 in the Dry Run (taken from MLD1).

temperature of the ULF at this level is also quite pronounced. The total values of the F_3 on this horizontal plane again almost entirely stem from the 3D ageostrophic velocity component. The location of large positive values also matches very well with the horizontal configuration of the ULF. There are negative values to the east of the disturbance center, suggesting that there is an ongoing frontolytical process there.

It is warranted to briefly compare the values of F_3 with the counterpart results of the traditional metric of frontogenetical function, F_2, in this simulation of ULF. Figure 7.22 shows that the four panels of F_2 have comparable values and are a thousand times smaller than those of F_{3T}, largely stemming from the fact that $\theta_z \gg |\nabla_2 \theta|$. More importantly, the distribution of F_{2T} has poor correlation with the location of ULF. Although F_{2G} has positive values and correlates fairly well with the lower part (between 650 hPa and 800 hPa) of the ULF, F_{2A} has negative values of comparable magnitude. To a great extent the values of F_{2A} cancel those of F_{2G},

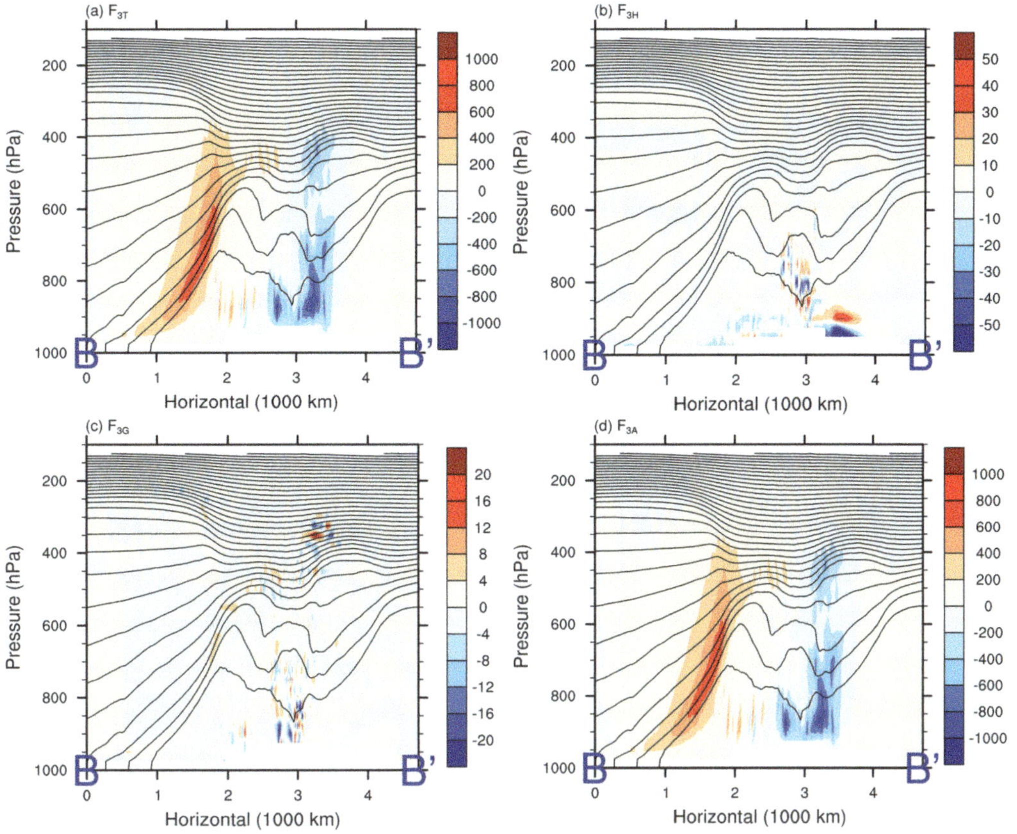

Figure 7.20. Distributions of F_{3T} and its three components F_{3H}, F_{3G}, F_{3A} (shade) in units of 10^{-8} K s^{-1} m^{-1} and potential temperature (contours) in K on the cross-section BB′ on day 6.5 in the Dry Run (taken from MLD1).

resulting in a poor correlation between the distribution of F_{2T} and the location of ULF in Figure 7.22a. We therefore conclude that F_2 is not an appropriate metric for the rate of development of ULF.

To check whether or not F_3 serves as a suitable measure of ULF throughout its development toward maturity, we plot in Figure 7.23 the corresponding results at 12 and 24 hours prior to day 6.5. Since the cyclone steadily propagates eastward as it intensifies in a zonally periodic domain, we have shifted all fields at these times by 93, 150, and 190 grid points to the west, respectively, so that the main feature of interest would appear at the center of each plotting domain on days 5.5, 6.0 and 6.5. Cross-section AA′ passing through the most pronounced part of the trough at the three times is practically the same *relative to* the moving trough; so also is cross-section BB′ passing through the ULF at an oblique angle. Panels (a) and (b) of Figure 7.23 are to be compared with Figure 7.15a; panels (c) and (d) of Figure 7.23 with Figure 7.19a; panels (e) and (f) of Figure 7.23 with Figure 7.20a. The high degree of similarity in

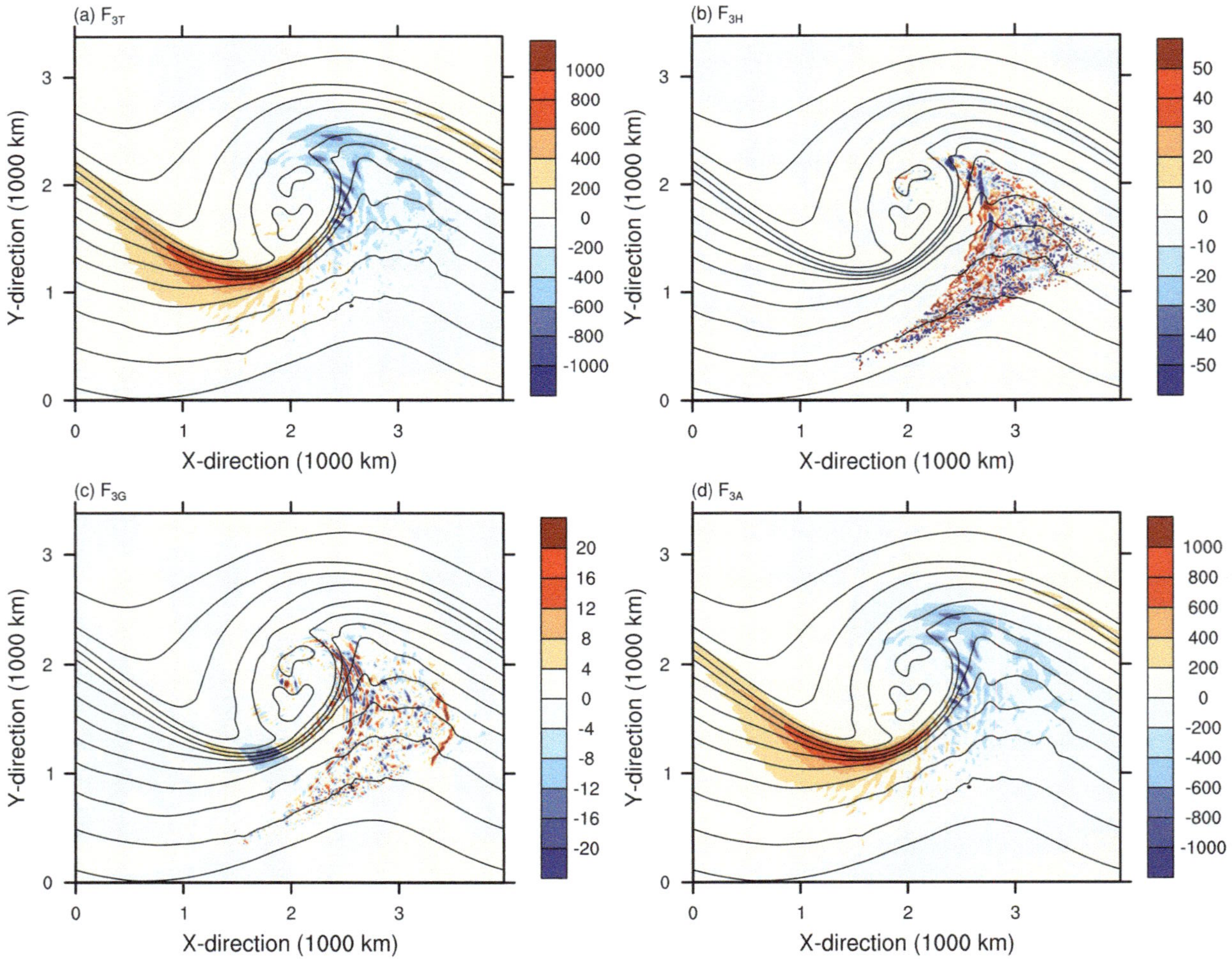

Figure 7.21. Horizontal distributions of F_3 and its three components (shading) in units of 10^{-8} K $s^{-1}m^{-1}$ and potential temperature (contours) in K at 700 hPa on day 6.5 in the Dry Run (taken from MLD1).

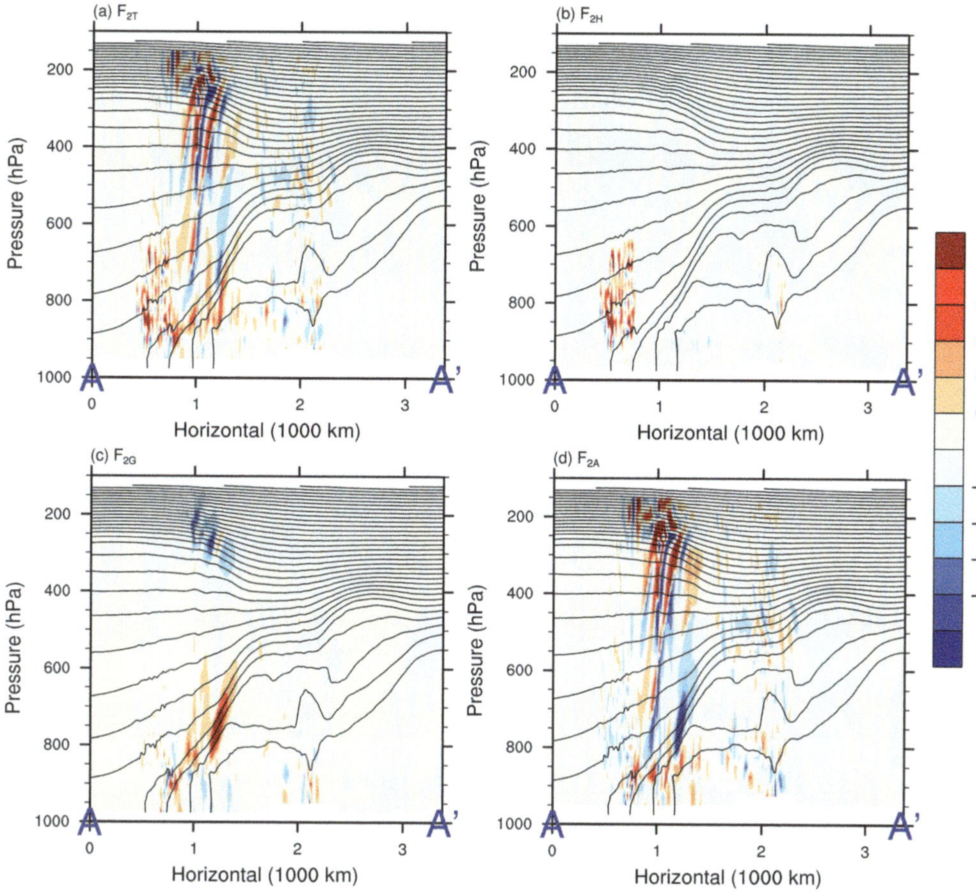

Figure 7.22. Vertical distributions of F_{2T} and its three components F_{2H}, F_{2G}, F_{2A} (shade) passing through AA′ in units of 10^{-8} K s^{-1} m^{-1} and potential temperature (contours) in K on day 6.5 in the Dry Run (taken from MLD1).

the counterpart results at these three times verifies that F_3 is indeed the appropriate metric throughout the development of ULF toward its maturity.

7.5.4 A Closer Examination of the 3D Frontogenetical Function

As we have done for F_{2G} and F_{2A} in Section 7.4.5, we now examine the contributions of the geostrophic and ageostrophic velocity components to F_{3G} and F_{3A} in terms of their deformation properties. We can write

$$F_{3G} \equiv -\frac{1}{|\nabla\theta|}\nabla_2\theta\cdot\boldsymbol{S} = -\frac{1}{|\nabla\theta|}\left[\theta_x\theta_y D_2 + \frac{1}{2}(\theta_x^2-\theta_y^2)D_1 + \hat{u}_z\theta_x + \hat{v}_z\theta_y\right] \qquad (7.11)$$

where $D_1 = \hat{u}_x - \hat{v}_y$ and $D_2 = \hat{u}_y + \hat{v}_x$ are the stretching deformation and shearing deformation of the geostrophic velocity on the (x–y) plane. Since hydrostatic balance holds very well as an approximation in the flow associated with the fronts, the

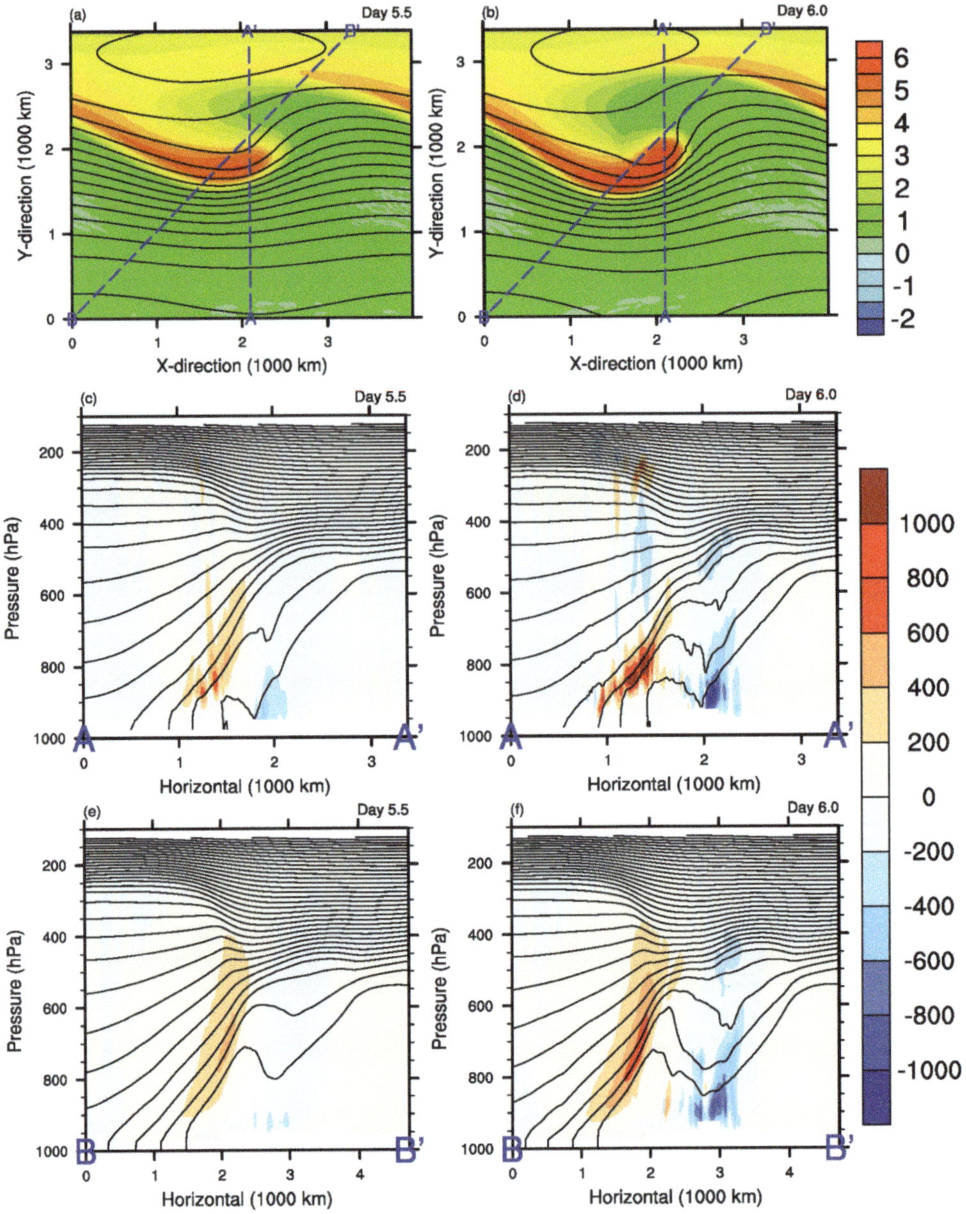

Figure 7.23. (a) Distributions of the potential vorticity (shading in PVU) and geopotential (contours) at 500 hPa on day 5.5 of the Dry Run; (c) and (e) distributions of potential temperature (contours) and $F_3 \equiv D|\nabla\theta|/Dt$ (shading; 10^{-8}K s^{-1} m^{-1}) at cross-sections AA′ and BB′ respectively; (b), (d) and (f) the counterparts of panels (a), (c) and (e) on day 6.0 respectively (taken from Mak et al., 2017).

thermal wind relation would apply with a high degree of accuracy, implying $(\hat{u}_{\tilde{z}}\theta_x + \hat{v}_{\tilde{z}}\theta_y) \approx 0$. Thus, F_{3G} entirely stems from the deformation property of the geostrophic velocity component $(\hat{u}, \hat{v})$. Recall that F_{3G} only accounts for a small percentage of F_3.

In a similar consideration of F_{3A}, we define the stretching and shearing deformation of $(\tilde{u}, \tilde{v}, \tilde{w})$ on the $(x$–$y)$ plane as $I_1 = \tilde{u}_x - \tilde{v}_y$ and $I_2 = \tilde{u}_y + \tilde{v}_x$ respectively; on the $(x$–$\tilde{z})$ plane as $J_1 = \tilde{u}_x - \tilde{w}_{\tilde{z}}$ and $J_2 = \tilde{u}_{\tilde{z}} + \tilde{w}_x$; and on the $(y$–$\tilde{z})$ plane as $K_1 = \tilde{v}_y - \tilde{w}_z$ and $K_2 = \tilde{v}_{\tilde{z}} + \tilde{w}_y$. Then F_{3A} can be rewritten as:

$$F_{3A} \equiv -\frac{1}{|\nabla\theta|}\nabla\theta \cdot A = M + N, \tag{7.12}$$

where

$$M = -\frac{1}{|\nabla\theta|}\left(\theta_x\theta_y I_2 + \theta_x\theta_{\tilde{z}} J_2 + \theta_y\theta_{\tilde{z}} K_2 + \theta_x^2 I_1 + \theta_y^2 K_1\right)$$

$$N = -|\nabla\theta|\tilde{w}_{\tilde{z}}.$$

The computation reveals that the values of M are considerably smaller than those of N in the main part of the ULF, and that the distribution of N approximates F_{3A} well (figures not shown for the sake of brevity). But M is not entirely negligible. In other words, the joint influence of the vortex stretching process and the deformation property of the 3D ageostrophic flow component are largely responsible for the development of ULF. To avoid misunderstanding, we should stress that the geostrophic component of the flow makes an important indirect and indispensable contribution through its central role in the intensification of the unstable baroclinic wave itself.

7.5.5 Transverse Circulation

The projection of the 3D ageostrophic velocity on the meridional vertical cross-section AA′ is shown in Figure 7.24. The $(\tilde{v}, \tilde{w})$ field gives us an overall view of the transverse circulation. The figure reveals that this projected ageostrophic circulation has a cellular form spanning the whole troposphere. Its descending branch brings down stratospheric air to form the ULF as remarked earlier.

7.5.6 Internal Gravity Waves in the Model Simulation

We have drawn attention to the relatively small-scale features in the distributions of F_3 and F_2 that extend through the troposphere into the stratosphere in the model (e.g., Figure 7.22). These features suggest the existence of meso-scale gravity waves in the jet–wave–front system. It is not surprising to find such features in a WRF simulation. After all, other models (e.g., Zhang, 2004; Plougonven and Snyder, 2007) had simulated different modes of this class of waves. Nevertheless, it calls for a closer diagnosis in order to ascertain that those small features are indeed gravity waves. We do so by examining the horizontal and vertical distributions of the horizontal divergence field. Figure 7.25 shows the horizontal distribution of that quantity together with the horizontal velocity field at 500 hPa on

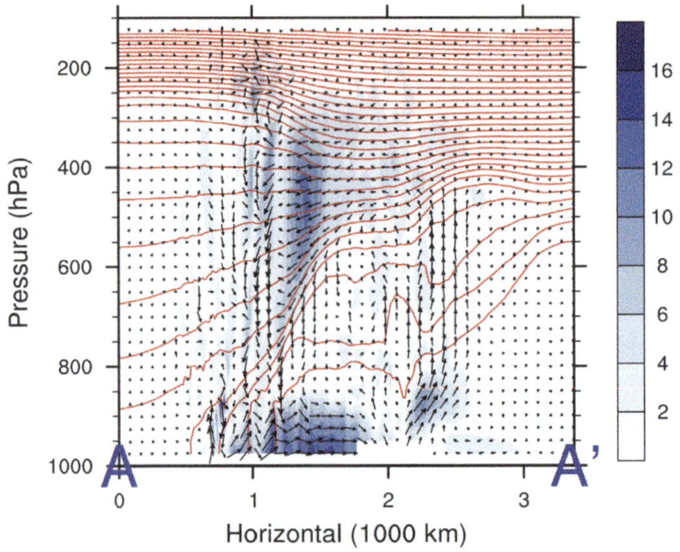

Figure 7.24. Distribution of the ageostrophic circulation projected on the AA′ cross-section, $(\tilde{v}, \tilde{\omega})$, together with the potential temperature on day 6.5 in the Dry Run; magnitude of $\tilde{\omega}$ is increased by a factor 1000 only for plotting (taken from MLD1).

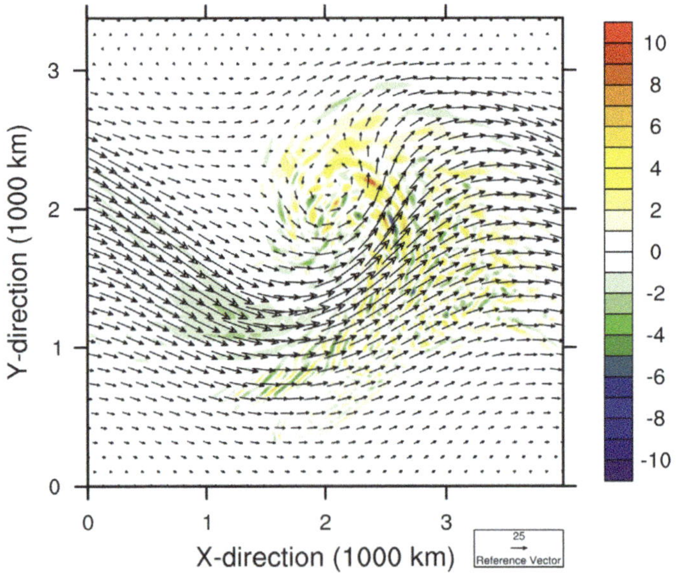

Figure 7.25. Distributions of the horizontal divergence in 10^{-5} s^{-1} (shading) and the horizontal velocity in m s^{-1} (vectors) at the 500 hPa level on day 6.5 of the Dry Run (taken from MLD1).

day 6.5 of the Dry Run. There is an extensive wavy pattern of convergence and divergence in the eastern half of the baroclinic wave trough. This is a clear signature of meso-scale gravity wave motions embedded in the baroclinic wave. We also see a large area of negative values slightly to the west of the trough in Figure 7.25.

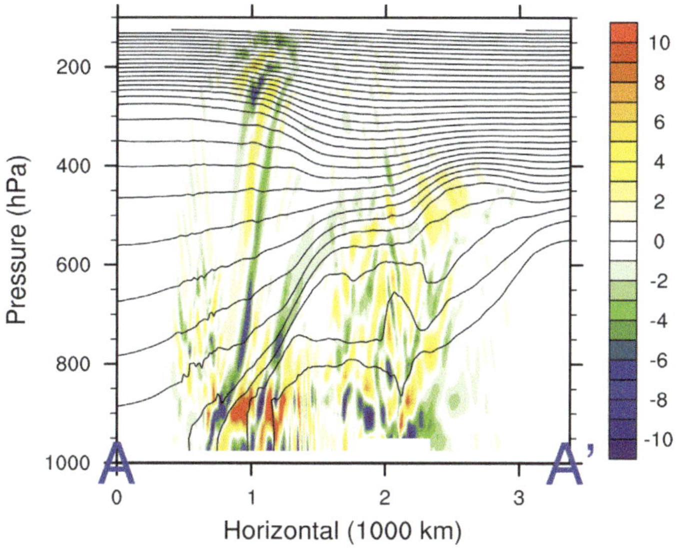

Figure 7.26. Distributions of the horizontal divergence in 10^{-5} s^{-1}(shading) and potential temperature (contours) on the cross-section passing through AA′ on day 6.5 of the Dry Run (taken from MLD1).

It is an area of horizontal convergence and hence corresponds to the location of ULF where there is synoptic-scale descent on the cross-section BB′, discussed earlier.

Figure 7.26 shows the vertical distribution of the horizontal divergence and the potential temperature on a cross-section passing through AA′ on day 6.5 of the Dry Run. The overall pattern can be interpreted as a superposition of several upward-propagating trains of meso-scale gravity waves. In addition, there is a distinct column of negative values collocated with the ULF. It indicates a layer of horizontal convergence associated with the descending motion seen in Figure 7.18a. Figures 7.25 and 7.26 together convincingly depict the presence of meso-scale gravity waves in our model.

7.5.7 Impacts of Diabatic Heating on the ULF

The diabatic heating in the Control Run is mainly due to condensational heating because the radiative time scale is relatively long. The initial distribution of the mixing ratio of moisture is zonally uniform, with a surface value of 16 g kg^{-1} at the southern boundary, decreasing northward to 1 g kg^{-1} in the domain. Vertically it is mostly confined to the lower troposphere. While there is considerable precipitation along a surface cold front on day 6.5, particularly abundant precipitation falls in the part of a warm front region near the center of the disturbance. The location of $x = 0$ has been shifted eastward by 200 grid points in the figures of this section. Figure 7.27 shows the model six-hour precipitation rate on day 6.5 in the Control

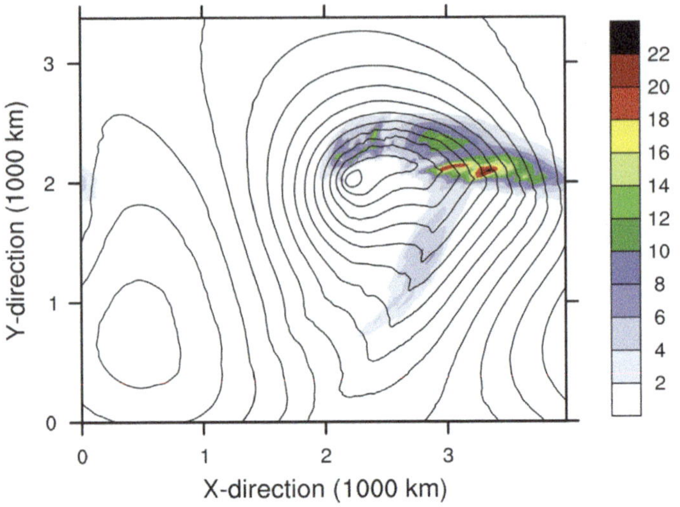

Figure 7.27. Distribution of six-hour precipitation rate in mm on day 6.5 in the Control Run (taken from MLD1).

Run, together with the surface pressure. The maximum value reaches 22 mm, which is a significant rate.

Not surprisingly, a ULF also develops in the model simulation with diabatic heating. It suffices to make a brief comparison of the properties of the ULF in the two simulations. Figure 7.28 shows the PV field on day 6.5 in the Control Run. It is to be compared with Figure 7.15 for the counterpart in the Dry Run. The baroclinic wave naturally intensifies more strongly in the Control Run and consequently the related fronts are stronger. We see that the baroclinic wave trough is deeper and the tongue of high PV values penetrates much further south than in the Dry Run.

The counterpart results of Figure 7.17 for the vertical distributions of PV are shown in Figure 7.29. It clearly depicts the parts of the ULF on the AA′ and BB′ cross-sections. The ULF lies along the leading edge of the narrow tongue of high PV values. In addition, there is a more pronounced layer of high PV values in the lower troposphere near the center of the disturbance, about 2000 km from point A as well as about 3000 km from point B. Such high PV value is mostly attributable to condensational heating, although to a lesser extent to the impact of turbulent mixing there.

The values of F_3 on the cross-section passing through BB′ in the Control Run are shown in Figure 7.30. The large values are again collocated nicely with ULF. We have used different color bars for the plotting of F_{3T}, F_{3H}, F_{3G} and F_{3A} in order to bring out the detailed features in each. The plots again confirm that F_3 is an appropriate measure of the rate of frontal development of a ULF in this case. It is noteworthy that the diabatic heating does not directly contribute to F_3 to a significant degree, but does so indirectly through its strong impact upon the ageostrophic velocity field of the intensifying baroclinic wave. The condensational heating gives rise to stronger small-scale features to the east of the ULF.

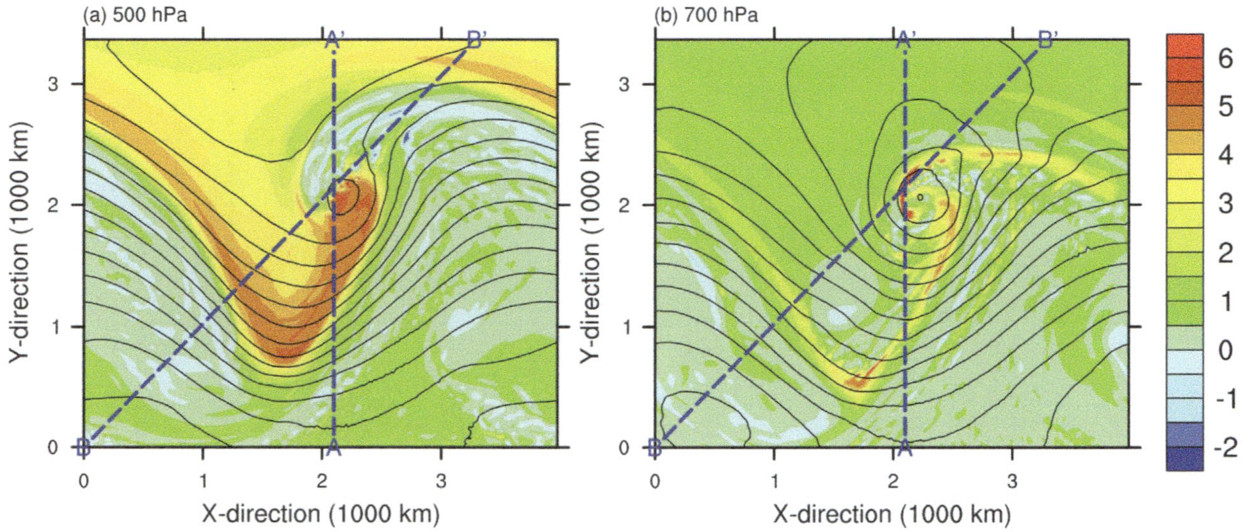

Figure 7.28. Distributions of PV (shading in PVU) and geopotential height (contours) at (a) 500 mb and (b) 700 mb levels on day 6.5 of the Control Run (taken from MLD1).

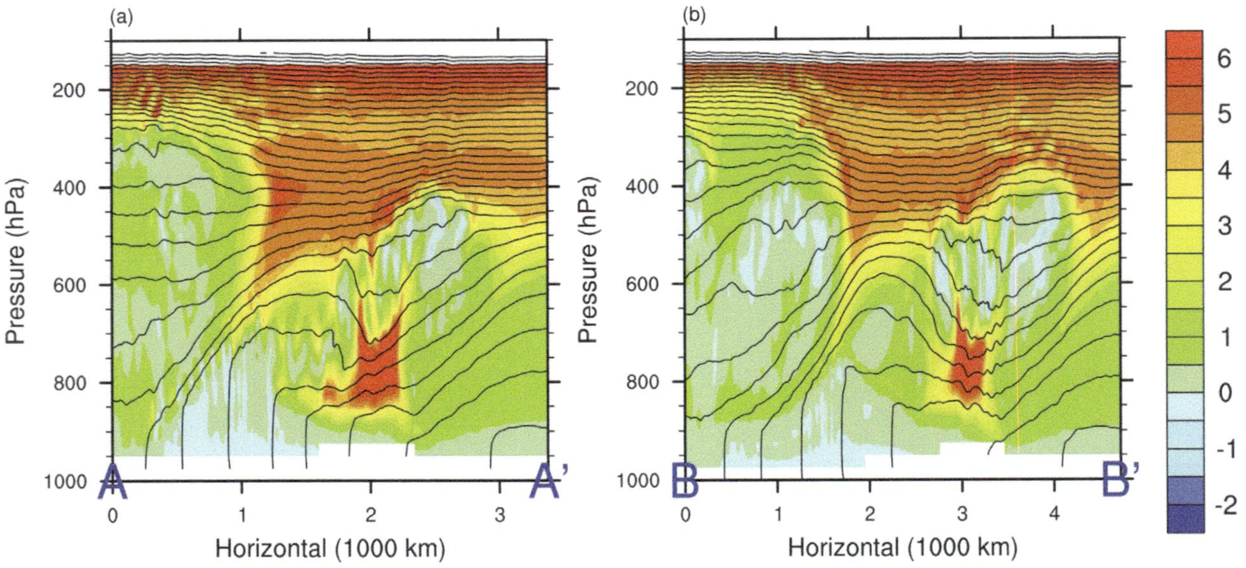

Figure 7.29. Distributions of PV in PVU (shading) and potential temperature in K (contours) on the vertical cross-section passing (a) through AA′ and (b) through BB′ on day 6.5 of the Control Run (taken from MLD1).

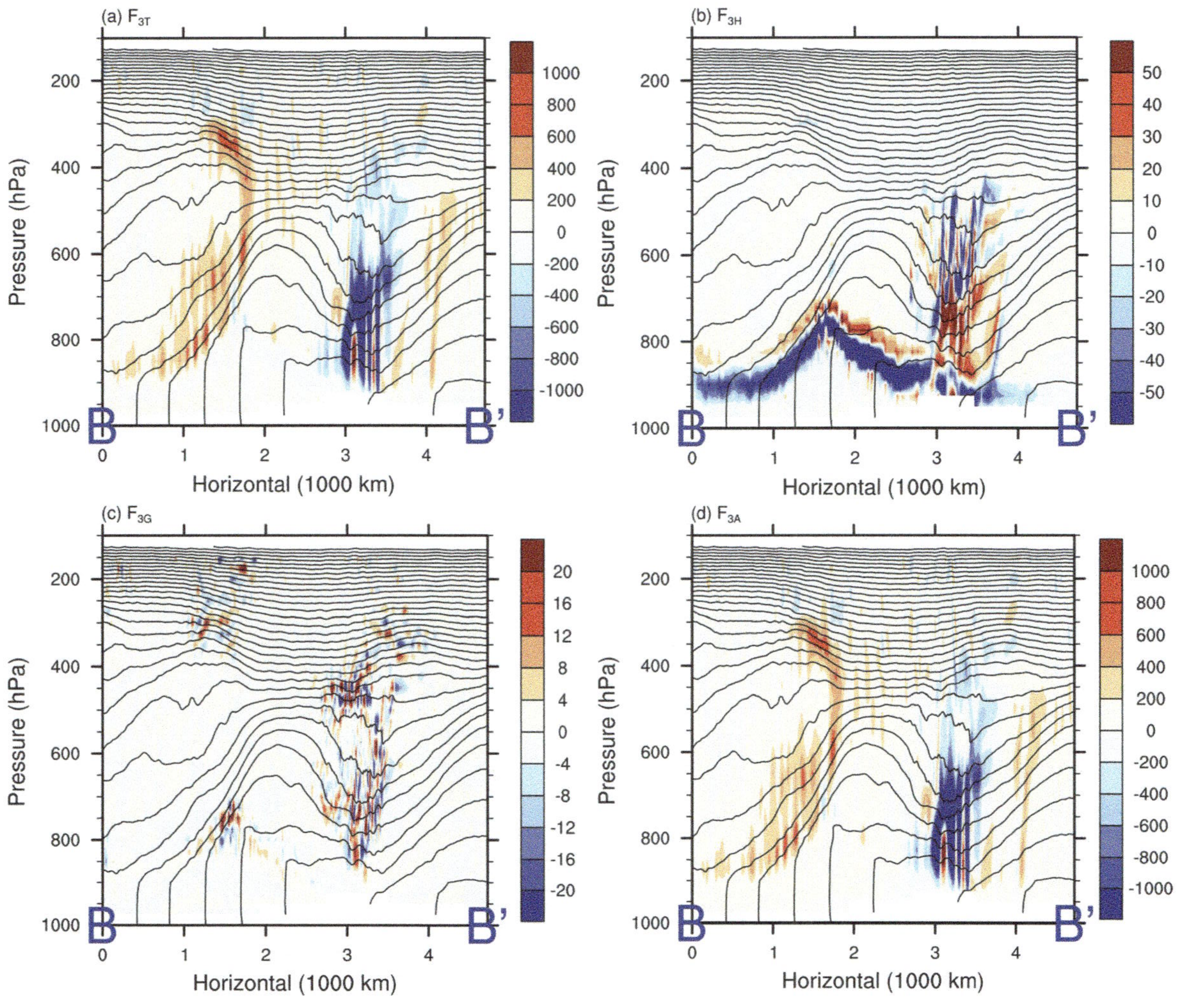

Figure 7.30. Distributions of F_3 and the three components (shading) in units of 10^{-8} K s^{-1} m^{-1} and potential temperature (contours) in K on the cross-section BB′ on day 6.5 in the Control Run (taken from MLD1).

7.6 Interaction Between the Upper-Level Front and Surface Fronts

We next address the questions: Do ULFs sometimes synergistically interact with surface fronts? If so, under what circumstance and why? Some observational studies concerning this issue have been cited and commented on in Section 7.1. We begin by reporting an additional example of such an interaction in an observed cyclone–front system.

7.6.1 Observed Interaction Between ULF and SF

Figure 1.1 (reproduced here as Figure 7.31) shows a swath of satellite image of MODIS (Moderate Resolution Imaging Spectroradiometer) taken at 1905Z on 26 September 2011, when it flew over the Midwestern United States. This disturbance has a gigantic comma-shaped cloud pattern made up of several types of cloud in different parts of it: innumerable low-level shallow cumulous clouds, organized vigorous convective cloud bands and high-level cirrus cloud decks. The comma pattern of clouds has a signature "tail" in the NE–SW orientation of about 1500 km. Overlaid onto this satellite image are contours of three meteorological variables constructed with the use of the ERA-Interim (ECMWF reanalysis dataset) at a slightly earlier time (1800Z). One of these fields is "two-meter temperature" (green contours). It increases by about 10 °C over a distance of 100 km across the most intense part of the signature "tail" in the cloud field. The curvilinear line at the western edge of it is a *cold front* separating surface warm moist air to the southeast from cold dry air to the northwest.

Figure 7.31 also shows the distributions of geopotential height (Z, black contours) and potential vorticity (PV, red contours) at the 425 hPa level. The lowest contour of Z is 6300 m, practically overhead of the minimum temperature center at the surface. In other words, this disturbance has developed an upper-level cutoff low and is in its mature stage by this time. Potential vorticity and potential temperature are quasi-conservative properties of air parcels if diabatic heating and turbulent mixing are both negligible. The PV in the stratosphere of the mean atmosphere is much larger than that in the troposphere, considerably larger than $1\text{PVU} = 10^{-6}\ \text{K s}^{-1}\text{kg}^{-1}\ \text{m}^2$. Suffice to note, there is an extensive area in Figure 7.31 enclosed by a contour of 1 PVU. The boundary of this area is highlighted by the heavy red contour. The maximum PV value within this enclosed area is equal to 4.6 PVU. The air in the enclosed area is therefore of stratospheric origin and must have been brought down to this tropospheric level through a *local tropopause folding*. This inference is consistent with the fact that the locations of the two centers of maximum PV values are distinctly less cloudy than their immediate surrounding. The boundary of this area is therefore a demarcation separating tropospheric air from air of stratospheric origin. As such, the heavy red contour is a *ULF*, which has the shape of a closed loop at this level and at this time. This ULF and the SF are therefore integral parts of a single extratropical cyclone, but have distinctly different properties.

A cross-section passing through the points indicated by M and N in Figure 7.31 would cut through the ULF and the surface cold front. Figure 7.32a shows the

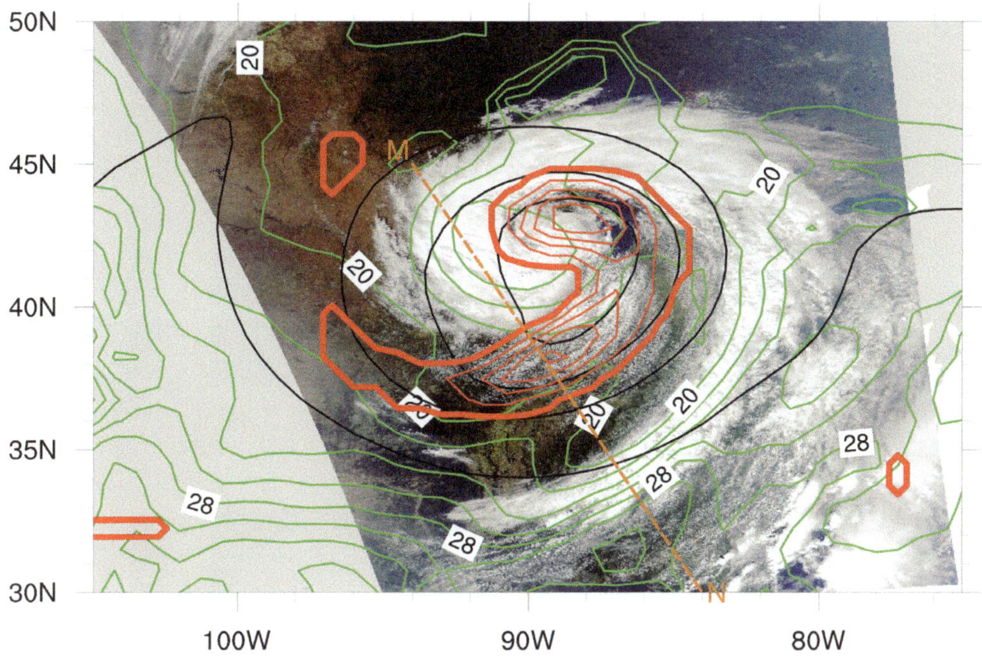

Figure 7.31. MODIS satellite image of a cyclone over the Midwestern United States at 1905Z on 26 September 2011. Three overlaid fields constructed with the ERA-Interim reanalysis data at 1800Z: (1) temperature at the 2 m level (green contours, T_{min} = 13 °C, CI = 2 °C); (2) potential vorticity at 425 hPa (red contours, $q_{\max} = 4.6$ PVU, CI = 1, thick red contour PV = 1 (1PVU = 10^{-6} K m^2 s^{-1}kg^{-1}); and (3) height of 425 hPa (black contours, $Z_{\min} = 6300$ m, CI = 100 m); orange dashed line MN indicates the location of a cross-section for further examination (taken from MLD2).

vertical distributions of the potential temperature as well as potential vorticity. The ULF links up with the surface cold front, suggesting that these two fronts plausibly interact in a mutually reinforcing manner. Figure 7.32b shows the distribution of the vertical velocity on the same cross-section. We see a pronounced cellular circulation linking the ULF to the surface cold front of the observed cyclone at 1200Z on 26 September 2011. The ascending motion over the surface cold front and descending motion at the ULF clearly portray a strong cellular circulation spanning the whole tropospheric depth. It would lead to significant synergistic interaction between the surface cold front and ULF. This quantitative evidence is deduced from arguably very reliable datasets. It supports the notion that ULF and SF associated with an intense mid-latitude cyclone likely reinforce one another.

7.6.2 Relevant Canonical Solutions of the Sawyer–Eliassen Equation

It is instructive to consider this dynamic problem in the context of the ageostrophic circulation governed by the Sawyer–Eliassen (SE) equation in the context of 2D SG dynamics. In Section 2.3.3 we discussed the canonical solutions of the SE

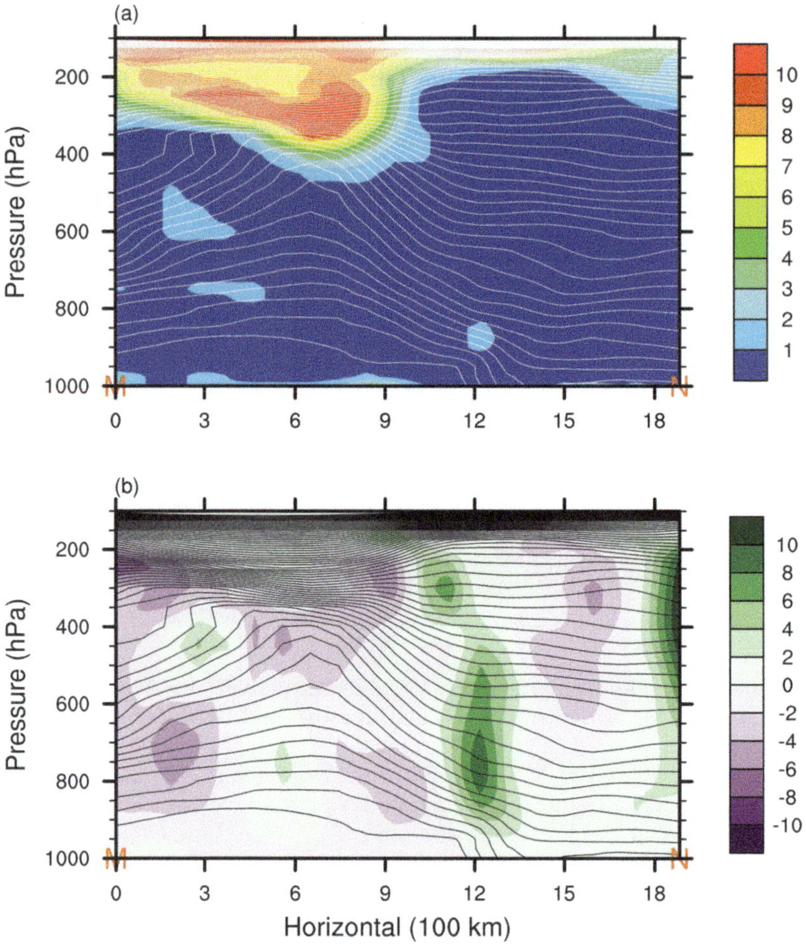

Figure 7.32. Distribution of (a) the potential vorticity in PVU (shading) and potential temperature (contours, CI = 2 K) on a vertical cross-section passing through the points M(45N, 94W) and N(30N, 84W) at 1200Z on 26 September 2011; and (b) the corresponding vertical velocity in units of cm s^{-1} (shading) and potential temperature (contours, CI = 2 K) (taken from MLD2).

equation under several idealized externally prescribed conditions (Hakim and Keyser, 2001). The responses in two situations are particularly instructive: a unit forcing at a mid-tropospheric level displaced either to the warm side or to the cold side of another unit forcing at a low level. Figure 7.33 shows the response on a cross-section normal to the *exit region* of an upper-level jet-streak overlying a low-level jet-streak. The thin dashed lines indicate the absolute momentum surfaces. Negative (positive) values of the ageostrophic streamfunction are indicated by dotted (solid) contours. Specifically, this transverse circulation is associated with a combination of a *negative* nondimensional unit forcing at the 5 km level and a *positive* nondimensional unit forcing at the 1 km level. The former is displaced 200 km to the *warm-air side* of the latter. The circulation associated with a negative

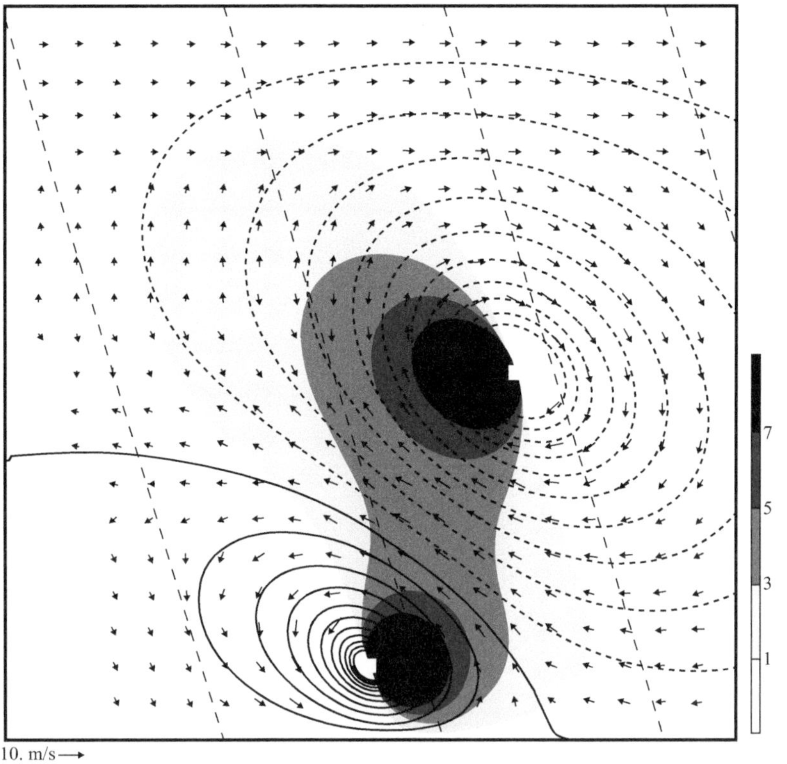

Figure 7.33. Semi-geostrophic solution for an upper-level jet-streak *exit region* overlying a surface frontal zone. Values of F^2, S^2 and N^2 coincide with those for the control case. Streamfunction is given by thick lines (negative values dashed) every 2×10^3 m^2 s^{-1}; positive values of vertical motion, w, are shaded every 2 cm s^{-1}, starting at 1 cm s^{-1}, and absolute momentum is given by thin dashed lines every 30 m s^{-1}. Vectors depict the ageostrophic circulation, with w scaled by a factor of 100 commensurate with the inverse aspect ratio of the physical dimensions of the domain, which is square in the plotting coordinates (reproduced from Hakim and Keyser, 2001).

(positive) unit forcing is a clockwise (anticlockwise) cell. Shading indicates the areas of ascent. The ascending motion reaches mid-tropospheric levels. The *positive interference* of these two cells results in having the ascent of one cell reinforcing that of the other cell. This is an example of vertical coupling. Notice that although the potential temperature field is not indicated in this figure, we can infer from the figure the locations of the fronts under consideration. The SF is a cold front located to the west of the area of ascending motion. The ULF is relatively weak, located about 300 km further to the east. The ULF is expected as a consequence of the descent that would bring down high PV air from further aloft. It should be emphasized that *the coupling of the two cells in this case should be described as a coupling of two jets, but* ***not*** *coupling of a ULF and a surface cold front.* When two jet-streaks interact, the related fronts may or may not interact (Shapiro, 1982). By the same

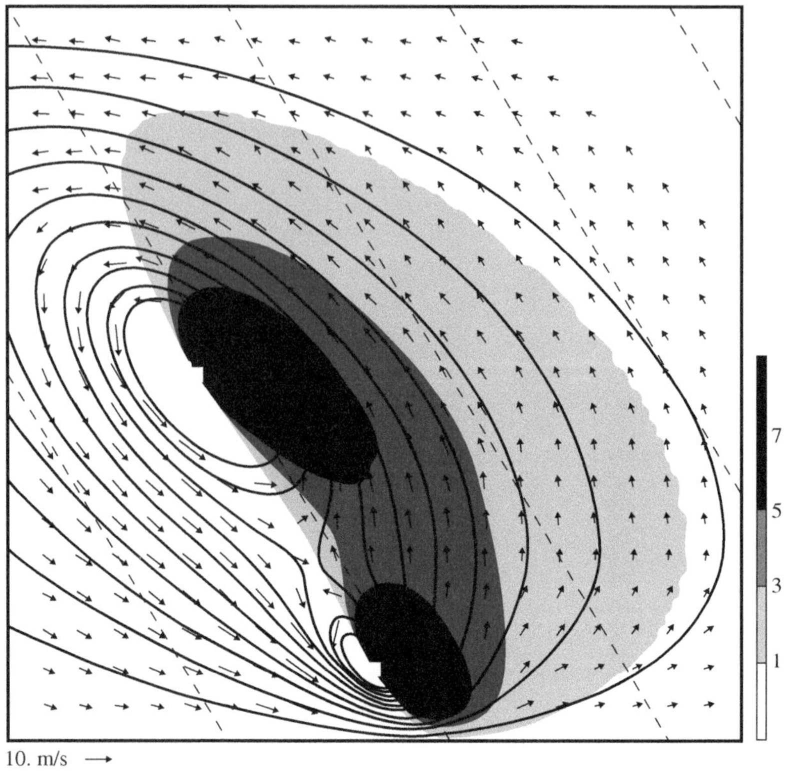

Figure 7.34. Semi-geostrophic solution for an upper-level jet-streak entrance region overlying a surface frontal zone. Streamfunction is given by thick lines (negative values dashed) every 2×10^3 m^2 s^{-1}; positive values of vertical motion w are shaded every 2 cm s^{-1} starting at 1 cm s^{-1}, and absolute momentum is given by thin dashed lines every 30 m s^{-1}. Vectors depict the ageostrophic circulation, with w scaled by a factor of 100 commensurate with the inverse aspect ratio of the physical dimensions of the domain, which is square in the plotting coordinates (reproduction of figure 8 in Hakim and Keyser, 2001).

token, if there is a negative upper-level forcing located to the cold side of a positive low-level forcing, there would be negative interference between the transverse cells and there would be no vertical coupling in such a case.

Figure 7.34 shows another canonical solution of the SE equation. This secondary circulation is on a cross-section normal to the *entrance region* of an upper-level jet-streak overlying a low-level jet-streak. In this case, there is a *positive* nondimensional unit forcing at the 5 km level and another *positive* nondimensional unit forcing at the 1 km level lying on a common absolute momentum surface. The former is displaced 240 km to the *cold-air-side* of the latter. There is *positive interference* between the two cells of transverse circulation. The coupling in this case is stronger than that of the previous case. The ascending motion (shaded area) of the lower cell reinforces that of the upper cell. The descending motion of the two cells reinforce likewise. The reinforced motion is effectively that of a single combined cell. The ascent extends to the upper levels on a plane essentially parallel to the absolute momentum surfaces. We can infer that the front at the surface is a pronounced cold front and there is a fairly strong ULF aloft to the west of the surface front. *This is an example of vertical coupling of not only the two jets, but also of a ULF and a surface cold front.* This idealized canonical solution of an SE equation tells us what it would take for a ULF to synergistically interact with an SF. In summary, a ULF and an SF may or may not interact. When they do, they may interact negatively or positively.

7.6.3 Dynamic Nature of Interaction Deduced from Model Simulation

Let us now examine the dynamics of vertical coupling of the ULF and SF on the basis of the model frontal simulation in the Control Run presented in the earlier sections. The distribution of potential vorticity q at 500 hPa on day 6.5 has been shown in Figure 7.28. Let us first make a more precise determination of the locations of the fronts under consideration. Computation of $|\nabla_2 q|$ is used to quantitatively ascertain the configuration of the ULF. Figures 7.35a, 7.35c and 7.35e show that $|\nabla_2 q|$ increases sharply across an almost continuous band with a maximum value of about 6 PVU per 100 km at the same three moments eight hours apart previously chosen for Figure 7.6. It establishes a highly curvilinear ULF in the trough region of the baroclinic wave separating air of tropospheric and stratospheric origins. We have also discussed before that this ULF extends down to different levels along it at this time.

To quantitatively delineate the location and intensity of surface cold and warm fronts, we likewise compute the distribution of $|\nabla_2 \theta|$ on the 950 hPa on day 6.5 (recall that the minimum surface pressure in the Control Run is 972 hPa). Figures 7.35b, 7.35d and 7.35f show such a result together with the surface wind (vectors) at the same three moments eight hours apart previously chosen for Figure 7.6. These panels quantify the evolution of the locations and widths of the SFs from the 4th to the 20th hour on day 6. The maximum value in the cold front with heating is about 25 K per 100 km. This intense cold front has a width of only about 50 km. The width of the cold front and the ULF are comparable. These panels again highlight the signature characteristic that the cold front is considerably longer, narrower, and stronger than the warm front, as we noted in Section 4.1. The band of maximum value of $|\nabla_2 \theta|$ at the center of the long and narrow cold front reaches 20 units. The values of the warm front are much less coherent. The surface wind is quite realistic, with a sharp turn in direction across the cold front. The wind is mostly southerly in the warm sector to the east of it. There is a more gradual veering across the broad warm front region. A short occluded front is evident between the low-pressure center and the intersection point of the cold and warm fronts. It is interesting to find a trace of a secondary cold front further west of the primary cold front in the Control Run. Such a feature seems to be

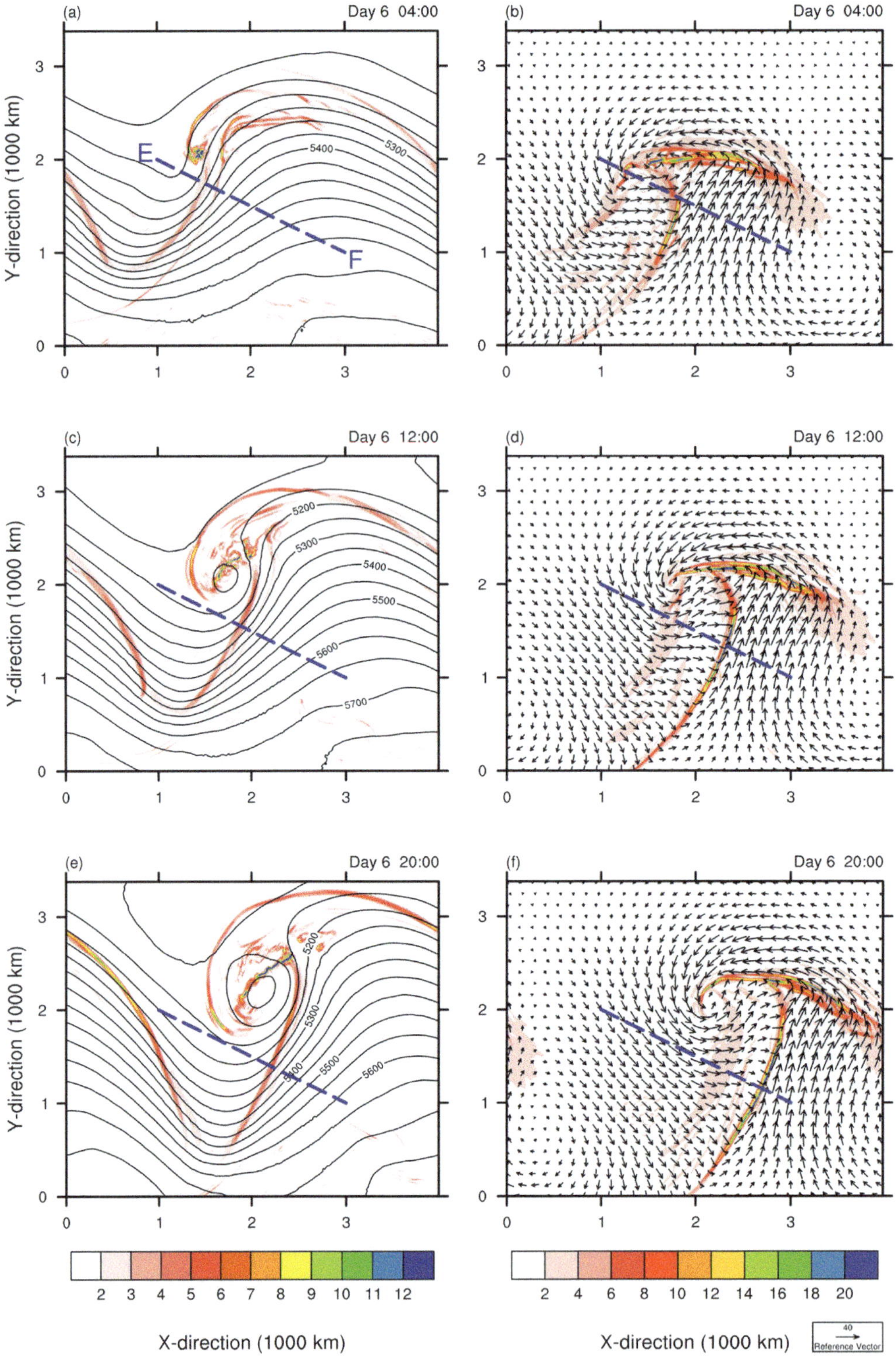

Figure 7.35. Distribution of the ULF and SF from 4th to 20th hour on day 6 in the Control Run: Left column, geopotential height of 500 hPa (contours) in meters and $|\nabla_2 q|$ (shading) in $10^{-11}(\mathrm{K\ m^2\ s^{-1}\ kg^{-1}})\mathrm{m}^{-1}$ where q is potential vorticity. Right column, horizontal wind (vectors) and $|\nabla_2 \theta|$ (shading) at 925 hPa in K per 100 km where θ is potential temperature (taken from MLD2).

fairly common, but the cause for it is quite subtle and uncertain. The model warm front with diabatic heating is also quite realistic.

It is noteworthy that the eastern segment of the ULF aloft is located about 400 km to the west of the surface cold front. The two fronts are roughly parallel to one another. To crudely take into account the effect of condensational heating under the assumption of pseudo-adiabatic approximation, we examine the distributions of equivalent potential temperature θ_e and equivalent potential vorticity $q_e = \frac{1}{\rho}(\zeta_{abs} \cdot \nabla\theta_e)$ where ζ_{abs} is absolute vorticity and ρ is air density. Figures 7.36a, 7.36c and 7.36e show the distributions of such properties on a vertical cross-section passing through the dashed line EF indicated in Figure 7.34. The plots reveal that the θ_e field substantially differs from the θ field only (not shown) at the low levels near the surface cold front, where there is abundant moisture, ascending motion and thereby condensational heating. It follows that the q_e field also only differs from the q field over the cold front and at the low levels to the east of it. The ULF progressively reaches down to the surface, linking up with the surface cold front. The plotting domain in these three panels is only 2000 km wide in the horizontal direction, whereby we can clearly show in greater detail what happens when the eastern segment of the ULF and the SF progressively get into relative positions for increasingly interacting with one another during this 16-hour interval.

The peak of the interaction is arguably captured in Figure 7.36d, which shows the corresponding distributions of the vertical velocity and θ_e. There is a strong narrow ascending branch (actually a maximum value of 40 cm s^{-1}) over the surface cold front and a broader descending branch (maximum 10 cm s^{-1}) along the sloping ULF. This is a strong cellular circulation connecting the model ULF and surface cold front. The model result is consistent with the observational finding presented in Figures 7.31 and 7.32. The configuration of the transverse circulation shown in Figure 7.36 is reminiscent of the special canonical solution in UK01 (Figure 7.34) under qualitatively comparable conditions, even though the model ULF and SF are by no means two-dimensional. It is noteworthy that the ULF and SF are roughly parallel to one another and the eastern segment of the ULF is located about 400 km to the west of the surface cold front. That is a favorable circumstance for positive interaction between a ULF and a surface cold front. The existence of the secondary cold front to the northwest of the primary cold front might arise from the interaction between the ULF and the SF as it is almost directly beneath the ULF. In conclusion, when a ULF is closely parallel to a surface cold front and is located a few hundred kilometers to the cold side of the SF, they would mutually reinforce one another through their ageostrophic circulation. We may expect strongly enhanced weather development as a result of such vertical coupling.

We further delineate the nature of interaction between a ULF and an SF on the basis of the equation(s) that relates the ageostrophic circulation to the geostrophic circulation. Models of different complexity would have different versions of such equation(s), as summarized in Table 7.1.

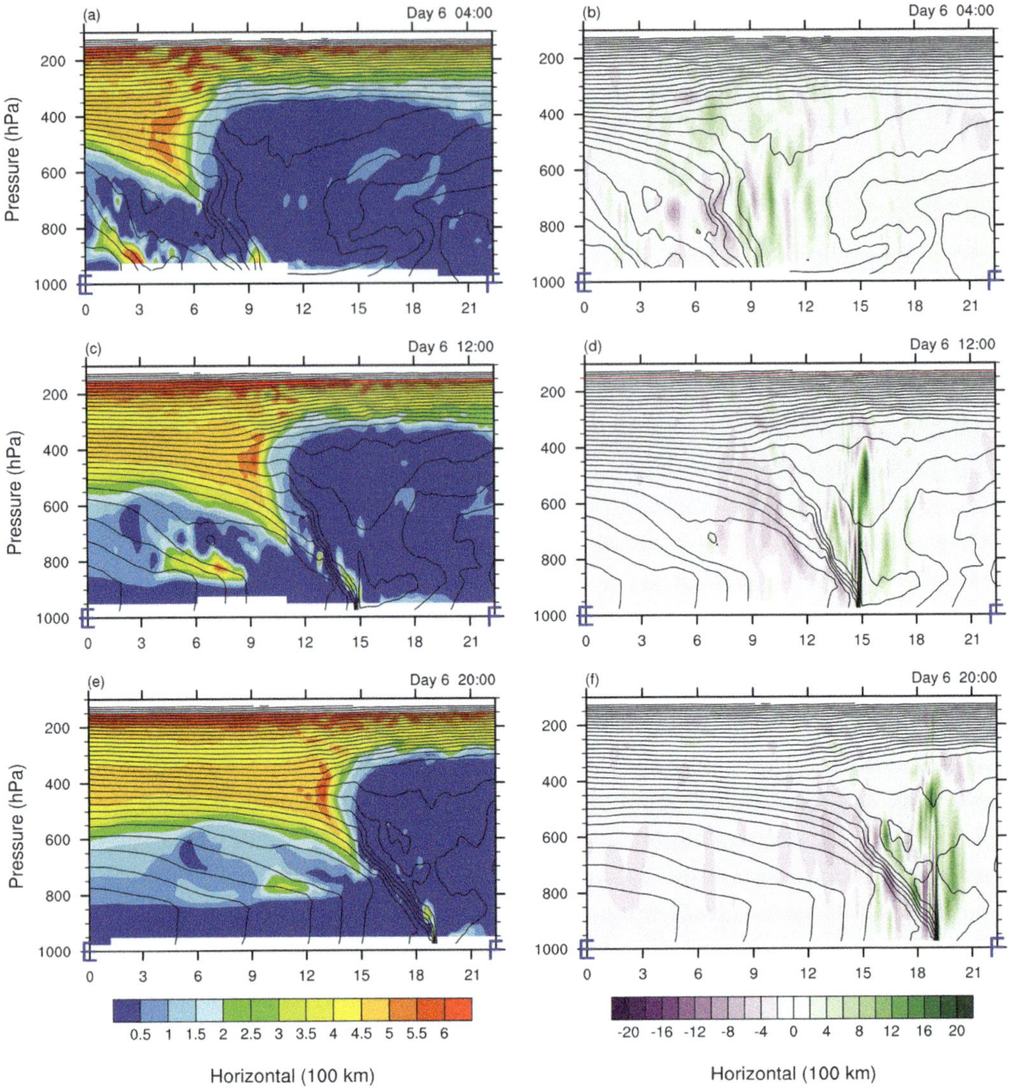

Figure 7.36. Left column, the equivalent potential vorticity in PVU (shading) and equivalent potential temperature in K (contours). Right column, the vertical velocity in cm s^{-1} (shading) and potential temperature distribution in K (contours, CI = 5K) on the cross-sections along the dashed lines EF as indicated in Figure 7.35a from the 4th to the 20th hour of day 6 in the Control Run (taken from MLD2).

It suffices to consider the gpSEs for a layer of rotating Boussinesq fluid with friction and heating in a 3D primitive equation model for the purpose of our discussion. A derivation of this is given in the Appendix. The resulting equations are (7.A11) and (7.A12) which are repeated below for the convenience of referencing:

$$S_1\{\eta\} + R_1\{\chi\} - \frac{\partial}{\partial z}\frac{D\tilde{v}}{Dt} = G + A\Gamma_x - F_z^{(y)} \tag{7.13}$$

Table 7.1. The four acronyms for the equations that relate the ageostrophic circulation to the geostrophic circulation

	Semi-geostrophic model	Primitive equation model
Two-dimensional	SEe	pSEe
Three-dimensional	gSEs	gpSEs

The four acronyms for the equations that relate the ageostrophic circulation to the geostrophic circulation in four progressively more general models are:

SEe: Sawyer–Eliassen equation (Sawyer, 1956; Eliassen, 1962);
pSEe: pseudo Sawyer–Eliassen equation (eq. (4.4) in Orlanski and Ross, 1977);
gSEs: generalized Sawyer–Eliassen system of equations (eqs. (10) and (11) in Mak, 2014);
gpSEs: generalized pseudo Sawyer–Eliassen system of equations (eqs. (7.13) and (7.14)).

$$R_2\{\eta\} + S_2\{\chi\} + \frac{\partial}{\partial z}\frac{D\tilde{u}}{Dt} = H + A\Gamma_y + F_z^{(x)}, \tag{7.14}$$

where $\left(F^{(x)}, F^{(y)}\right)$ is the frictional force per unit mass, Γ the diabatic heating rate in K s^{-1} and $A = \frac{g}{f\theta_{oo}}$ (all notations are defined in the Appendix). The three terms on the RHS of (7.13) and (7.14) represent three distinct types of "forcing" for the ageostrophic circulation. The fundamental differences between gpSEs and gSEs stem from the existence of the terms $\frac{\partial}{\partial z}\frac{D\tilde{u}}{Dt}$ and $-\frac{\partial}{\partial z}\frac{D\tilde{v}}{Dt}$, which can be explicitly written as

$$-\frac{\partial}{\partial z}\frac{D}{Dt}\tilde{v} = \frac{\partial}{\partial t}\chi_{zz} + L\{\chi\} + M\{\eta,\chi\} \tag{7.15}$$

$$\frac{\partial}{\partial z}\frac{D}{Dt}\tilde{u} = -\frac{\partial}{\partial t}\eta_{zz} - L\{\eta\} + N\{\eta,\chi\}. \tag{7.16}$$

$L = \hat{u}\frac{\partial^2}{\partial x\partial z} + \hat{v}\frac{\partial^2}{\partial y\partial z} + \hat{u}_z\frac{\partial}{\partial x} + \hat{v}_z\frac{\partial}{\partial y}$ is a linear differential operator, whereas M and N are nonlinear operators; $M\{\eta,\chi\} \equiv -\left[\tilde{u}\tilde{v}_{xz} + \tilde{v}\tilde{v}_{yz} + \tilde{w}\tilde{v}_{zz} + \tilde{u}_z\tilde{v}_x + \tilde{v}_z\tilde{v}_y + \tilde{w}_z\tilde{v}_z\right]$, $N\{\eta,\chi\} \equiv \left[\tilde{u}\tilde{u}_{xz} + \tilde{v}\tilde{u}_{yz} + \tilde{w}\tilde{u}_{zz} + \tilde{u}_z\tilde{u}_x + \tilde{v}_z\tilde{u}_y + \tilde{w}_z\tilde{u}_z\right]$. In light of (7.15) and (7.16) the gpSEs are prognostic rather than diagnostic in nature. Hence we refer to such equation(s) as "pseudo SE" in a PE model. Even under the assumption that the time-derivative terms could be neglected, the M and N nonlinear terms would make the task of accurately solving (7.13) and (7.14) very difficult.

Actually, there is no need to deduce the 3D ageostrophic circulation with the use of the gpSEs since it can be readily deduced from the WRF model output of the other dependent variables. In light of the nonlinear operators M and N, the resulting ageostrophic flows due to the individual types of forcing are not superposable. From a simplified diagnostic point of view, one might consider evaluating an approximate

3D ageostrophic circulation associated with each forcing under the assumption that the $\frac{\partial}{\partial z}\frac{D\tilde{u}}{Dt}$ and $-\frac{\partial}{\partial z}\frac{D\tilde{v}}{Dt}$ terms are sufficiently small. In principle, applying the methodology of Mak (2014) can do it. But it is not a technically feasible task because the related matrices in this model are extremely large $(m \times m, m = 263 \times 534 \times 16)$. We therefore limit ourselves to determining the projection of each type of forcing as well as the model ageostrophic circulation on a suitably chosen 2D vertical cross-section during frontogenesis. That would reveal the relative importance of the different forcing and their possible roles in influencing the ageostrophic circulation.

The forcing associated with the geostrophic circulation is represented by the geostrophic Q-vector, which is a horizontal vector at every grid point,

$$\vec{Q} = (G, H) = -2A\left(\hat{\theta}_x\hat{u}_x + \hat{\theta}_y\hat{v}_x,\ \hat{\theta}_x\hat{u}_y + \hat{\theta}_y\hat{v}_y\right).$$

The selected cross-section is to be approximately normal to both ULF and SF under consideration, such as the one passing through EF indicated in Figure 7.35c on the 12th hour of day 6. The horizontal coordinates of the endpoints of EF are $(x, y) = (1, 2)$ and $(x, y) = (3, 1)$ in units of 1000 km. A horizontal vector $\overrightarrow{EF}$ indicates the orientation of this plane. A horizontal unit vector on this cross-section is $\vec{I} = \overrightarrow{EF}/\left|\overrightarrow{EF}\right| = \frac{1}{\sqrt{5}}(2, -1)$. The unit vectors at all points are identical. The projection of the Q-vector at every point on such a selected cross-section is $\left(\vec{Q}\cdot\vec{I}\right)$. This is a 2D measure of such forcing for the ageostrophic circulation on this vertical cross-section. According to (7.13) and (7.14), the forcing arising from the diabatic heating processes at each grid point is represented by a vector $\vec{J} \equiv (A\Gamma_x, A\Gamma_y)$. It follows that the distribution of this forcing is $\left(\vec{J}\cdot\vec{I}\right)$. The frictional force due to turbulent mixing and diffusion in the WRF model is calculated in a boundary layer with the Yonsei University (YSU) PBL scheme (Hong et al. 2006).

The results of the three types of forcing in units of $10^{-6}\,\mathrm{s}^{-2}$ are shown in Figure 7.37a,b,c. The color bar of each panel is chosen to optimally show the characteristics of the forcing so that the extreme values would not excessively dominate the overall appearance of the plots. Figure 7.37a reveals significant positive values (around five units) of the forcing associated with the geostrophic circulation, $\vec{Q}\cdot\vec{I}$, all along the slope that links the ULF and the SF. However, the values are locally large in a very small area next to the surface front with maximum and minimum values of 42.5 and (–26.5) units respectively. Figure 7.37b shows values of generally larger magnitude in the forcing associated with the heating, $\vec{J}\cdot\vec{I}$, over a broader area above the SF extending up to 600 hPa. The extreme values are 26.0 and (–34.3) units respectively. Figure 7.37c shows much smaller values of the forcing associated with the frictional process in a boundary layer. The extreme values are only 5.3 and (–5.5) units respectively. The relative importance of the three types of forcing is therefore well documented by these three panels. Their distributions are quite different and complex.

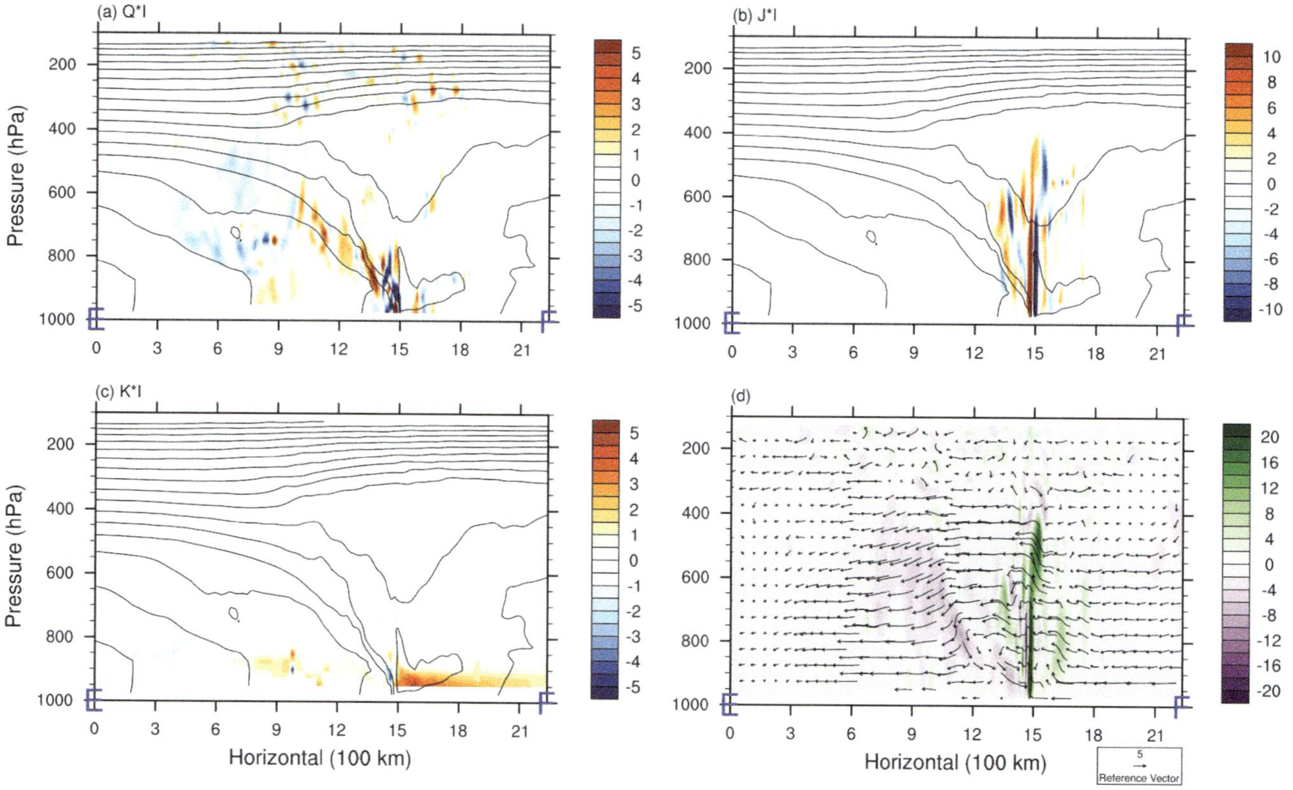

Figure 7.37. Distribution of the forcing for the ageostrophic circulation in units of $10^{-6}\ \mathrm{s}^{-2}$ on day 6.5 on the cross-section passing through EF as indicated in Figure 7.11c associated with the (a) geostrophic circulation, $\vec{Q}\cdot\vec{I}$, (b) diabatic heating $\vec{J}\cdot\vec{I}$, and (c) frictional force, $\vec{K}\cdot\vec{I}$; (d) distribution of $\tilde{w}$ in cms^{-1} and the projection of the 3D ageostrophic velocity on the same vertical cross-section, $(\mathfrak{H},\tilde{w})$ with $\mathfrak{H}=\vec{h}\cdot\vec{I}$, $\vec{h}=(\tilde{u},\tilde{v})$ with $\tilde{w}$ enhanced by a factor of 100 (taken from MLD2).

We have also determined the projection of the 3D ageostrophic velocity $(\tilde{u}, \tilde{v}, \tilde{w})$ on the cross-section under consideration. It is shown as a vector plot in Figure 7.37d, namely $(\mathfrak{H}\tilde{w})$ where $\mathfrak{H} = \vec{h} \cdot \vec{I}$, $\vec{h} = (\tilde{u}, \tilde{v})$ with $\tilde{w}$ amplified by a factor of 100. It portrays a broad cellular circulation linking the ULF and the SF that reflects the influence of the net forcing. In particular, the complicated detailed features over the SF are evidently associated with the diabatic heating seen in Figure 7.37b.

7.7 Summary Remarks

We summarize the main contents of this rather long chapter in a narrative format. It is confirmed that the rate of development of SFs can be adequately quantified with a 2D frontogenetic function F_2. It is found that cold fronts are characteristically longer, narrower, and more intense than warm fronts in the extratropical atmosphere. The differential thermal advection by the geostrophic and ageostrophic wind components in the two frontal regions results in a greater thermal contrast across the cold front. The length of the cold front is essentially the length scale of the intensifying baroclinic wave (i.e., of the order of radius of deformation). The frontal system as a whole moves eastward under the influence of a steering flow associated with the initial reference flow. In addition, the cold front outpaces the warm front eastward, making the western portion of the warm front progressively occluded and the eastern portion of the warm front shorter. The dynamical processes tend to move the cold front *eastward*, whereas the diabatic heating processes tend to move it *westward*, contributing to the narrowness of the cold front.

Tropopause folding in our simulation manifests as a ULF in which the local values of 3D potential temperature gradient are large. The upper-level synoptic-scale descent brings down stratospheric air of high PV values to different levels of the troposphere along the model ULF. The ULF is therefore a 2D surface that separates air of tropospheric and stratospheric origins. It would be meaningful to metaphorically think of the ULF as a "veil in the sky" that has varying thickness across the mid-section upstream of the trough of a baroclinic wave. It has been demonstrated that the 3D frontogenetical function F_3 is a necessary and sufficient metric for measuring the rate of development of a ULF. The diabatic heating processes do not directly contribute to F_3 to a significant degree, but they do so indirectly through their influence on the strength and structure of the subsidence during the intensification of the baroclinic wave itself. Furthermore, on the basis of the distribution of the horizontal divergence field, we verify that the jet–wave–front system generates a spectrum of meso-scale gravity waves that propagate into the stratosphere.

We have provided observational as well as model evidences in support of the notion that an intense ULF and an SF would synergistically reinforce one another when the ULF aloft is located several hundred kilometers to the west of the surface cold front. The transverse circulations associated with these two fronts would under this condition form a single mutually reinforcing westward-tilting cell that spans the whole troposphere. Such interaction would lead to especially severe weather events and great exchange of mass between the troposphere and stratosphere. It warrants making systematic statistical diagnoses to establish how commonly such interaction occurs in the atmosphere.

Appendix: Generalized Pseudo Sawyer–Eliassen System of Equations

This system of equations (gpSEs for short) relates the ageostrophic circulation to the geostrophic circulation in a 3D PE model. We derive such equations in this appendix for a layer of rotating Boussinesq fluid with the inclusion of frictional and heating processes. We may decompose the dependent variables as follows,

$$\begin{aligned} \text{Velocity}(u,\ v,\ w) &= (\hat{u} + \tilde{u},\ \hat{v} + \tilde{v},\ \tilde{w}) \\ \text{Potential temperature } \theta &= \Theta(z) + \hat{\theta} \\ \text{Pressure divided by density } \phi &= \Phi(z) + \hat{\phi}, \end{aligned} \tag{7.A1}$$

where $\hat{V} \equiv (\hat{u}, \hat{v})$ is total geostrophic velocity; $(\tilde{u}, \tilde{v}, \tilde{w})$, ageostrophic velocity; Θ and Φ, properties of a height-varying ambient component of the thermodynamic state. The auxiliary relations among $\hat{u}$, $\hat{v}$, $\hat{\phi}$ and $\hat{\theta}$ are geostrophic balance and hydrostatic balance by definition:

$$f\hat{v} = \hat{\phi}_x, \tag{7.A2a}$$

$$f\hat{u} = -\hat{\phi}_y, \tag{7.A2b}$$

$$\hat{\phi}_z = \frac{g}{\theta_{oo}}\hat{\theta}. \tag{7.A2c}$$

We denote the frictional force per unit mass by $\left(F^{(x)}, F^{(y)}\right)$ and the diabatic heating rate in K s^{-1} by Γ. They may be regarded as unspecified functions of $\hat{u}, \hat{v}$ and $\hat{\theta}$. The complete set of governing equations for this PE model is then

$$\hat{v}_z = A\hat{\theta}_x, \tag{7.A3a}$$

$$\hat{u}_z = -A\hat{\theta}_y, \tag{7.A3b}$$

$$\frac{D}{Dt}(\hat{u} + \tilde{u}) = f\tilde{v} + F^{(x)}, \tag{7.A4}$$

$$\frac{D}{Dt}(\hat{v} + \tilde{v}) = -f\tilde{u} + F^{(y)}, \tag{7.A5}$$

$$\frac{D\hat{\theta}}{Dt} + \tilde{w}\Theta_z = \Gamma, \tag{7.A6}$$

$$\tilde{u}_x + \tilde{v}_y + \tilde{w}_z = 0, \tag{7.A7}$$

where $A = \frac{g}{f\theta_{oo}}$ and $\frac{D}{Dt} = \frac{\partial}{\partial t} + (\hat{u} + \hat{u})\frac{\partial}{\partial x} + (\hat{v} + \hat{v})\frac{\partial}{\partial y} + (\tilde{w})\frac{\partial}{\partial z}$.

Combining the x-derivative of (7.A6) with the z-derivative of (7.A5) and making use of (7.A3a) would yield:

$$A_1\tilde{u}+B_1\tilde{v}+C_1\tilde{w}-\frac{\partial}{\partial z}\frac{D\tilde{v}}{Dt}=G+A\Gamma_x-F_z^{(y)}, \tag{7.A8}$$

where

$$A_1=\hat{v}_z\frac{\partial}{\partial x}-(f+\hat{v}_x)\frac{\partial}{\partial z},\ B_1=-\hat{u}_z\frac{\partial}{\partial x}-\hat{v}_y\frac{\partial}{\partial z},\ C_1=A\left(\Theta_z+\hat{\theta}_z\right)\frac{\partial}{\partial x}-\hat{v}_z\frac{\partial}{\partial z},$$

$$G=2\hat{u}_z\hat{v}_x+2\hat{v}_y\hat{v}_z=-2A\nabla\hat{\theta}\cdot\frac{\partial}{\partial x}\hat{V},\ \nabla\equiv\left(\frac{\partial}{\partial x},\frac{\partial}{\partial y}\right).$$

Combining the y-derivative of (7.A6) with the z-derivative of (7.A4) and making use of (7.A3b) would likewise yield:

$$A_2\tilde{u}+B_2\tilde{v}+C_2\tilde{w}+\frac{\partial}{\partial z}\frac{D\tilde{u}}{Dt}=H+A\Gamma_y+F_z^{(x)}, \tag{7.A9}$$

where:

$$A_2=\hat{v}_z\frac{\partial}{\partial y}+\hat{u}_x\frac{\partial}{\partial z},\ B_2=-\hat{u}_z\frac{\partial}{\partial y}-(f-\hat{u}_y)\frac{\partial}{\partial z},\ C_2=A\left(\Theta_z+\hat{\theta}_z\right)\frac{\partial}{\partial y}+\hat{u}_z\frac{\partial}{\partial z},$$

$$H=-2\hat{v}_z\hat{u}_y-2\hat{u}_z\hat{u}_x=-2A\nabla\hat{\theta}\cdot\frac{\partial}{\partial y}\hat{V}.$$

The next step is to define $\tilde{w}=\tilde{w}^{(1)}+\tilde{w}^{(2)}$ with $\tilde{u}_x+\tilde{w}_z^{(1)}=0$ and $\tilde{v}_y+\tilde{w}_z^{(2)}=0$ in order to satisfy (7.A7). It follows that we may introduce two ageostrophic potentials η and χ such that:

$$\tilde{u}=-\eta_z,\ \tilde{v}=-\chi_z,\ \tilde{w}=\eta_x+\chi_y. \tag{7.A10}$$

Upon substituting the expressions in (7.A10) into (7.A8) and (7.A9), we finally get

$$S_1\{\eta\}+R_1\{\chi\}-\frac{\partial}{\partial z}\frac{D\tilde{v}}{Dt}=G+A\Gamma_x-F_z^{(y)}, \tag{7.A11}$$

$$R_2\{\eta\}+S_2\{\chi\}+\frac{\partial}{\partial z}\frac{D\tilde{u}}{Dt}=H+A\Gamma_y+F_z^{(x)}, \tag{7.A12}$$

where

$$S_1=(f+\hat{v}_x)\frac{\partial^2}{\partial z^2}-2\hat{v}_z\frac{\partial^2}{\partial x\partial z}+A\left(\Theta_z+\hat{\theta}_z\right)\frac{\partial^2}{\partial x^2},$$

$$R_1=\hat{v}_y\frac{\partial^2}{\partial z^2}+A\left(\Theta_z+\hat{\theta}_z\right)\frac{\partial^2}{\partial x\partial y}-\hat{v}_z\frac{\partial^2}{\partial y\partial z}+\hat{u}_z\frac{\partial^2}{\partial x\partial z},$$

$$S_2 = (f - \hat{u}_y)\frac{\partial^2}{\partial z^2} + 2\hat{u}_z\frac{\partial^2}{\partial y \partial z} + A(\Theta_z + \hat{\theta}_z)\frac{\partial^2}{\partial y^2},$$

$$R_2 = -\hat{u}_x\frac{\partial^2}{\partial z^2} + A(\Theta_z + \hat{\theta}_z)\frac{\partial^2}{\partial x \partial y} - \hat{v}_z\frac{\partial^2}{\partial y \partial z} + \hat{u}_z\frac{\partial^2}{\partial x \partial z}.$$

G and H are functions of the geostrophic velocity. The vector (G, H) is also known as the geostrophic **Q**-vector, representing a forcing associated with the geostrophic circulation. Two vectors, $(A\Gamma_x, A\Gamma_y)$ and $(-F_z^{(y)}, F_z^{(x)})$, represent the influence of heating and friction respectively. Equations (7.A11) and (7.A12) are the gpSEs for this model. If we drop the $\frac{\partial}{\partial z}\frac{D\tilde{u}}{Dt}$ and $-\frac{\partial}{\partial z}\frac{D\tilde{v}}{Dt}$ terms, the equations would be reduced to eqs. (10) and (11) of Mak (2014) for a 3D SG model, albeit they include the additional effects of friction and heating here.

The fundamental differences between gpSEs and gSEs stem from the existence of the terms $\frac{\partial}{\partial z}\frac{D\tilde{u}}{Dt}$ and $-\frac{\partial}{\partial z}\frac{D\tilde{v}}{Dt}$, which can be explicitly written as:

$$-\frac{\partial}{\partial z}\frac{D}{Dt}\tilde{v} = \frac{\partial}{\partial t}\chi_{zz} + L\{\chi\} + M\{\eta, \chi\}, \tag{7.A13}$$

$$\frac{\partial}{\partial z}\frac{D}{Dt}\tilde{u} = -\frac{\partial}{\partial t}\eta_{zz} - L\{\eta\} + N\{\eta, \chi\}. \tag{7.A14}$$

$\mathrm{L} = \hat{u}\frac{\partial^2}{\partial x \partial z} + \hat{v}\frac{\partial^2}{\partial y \partial z} + \hat{u}_z\frac{\partial}{\partial x} + \hat{v}_z\frac{\partial}{\partial y}$ is a linear differential operator, whereas M and N are nonlinear operators: $M\{\eta, \chi\} \equiv -[\tilde{u}\tilde{v}_{xz} + \tilde{v}\tilde{v}_{yz} + \tilde{w}\tilde{v}_{zz} + \tilde{u}_z\tilde{v}_x + \tilde{v}_z\tilde{v}_y + \tilde{w}_z\tilde{v}_z]$, $N\{\eta, \chi\} \equiv [\tilde{u}\tilde{u}_{xz} + \tilde{v}u_{yz} + \tilde{w}\tilde{u}_{zz} + \tilde{u}_z\tilde{u}_x + \tilde{v}_z\tilde{u}_y + \tilde{w}_z\tilde{u}_z]$. Even under the assumption that the time derivative terms in (7.A13) and (7.A14) could be ignored, the additional nonlinear terms would make the task of accurately solving (7.A11) and (7.A12) very difficult, if at all possible.

Since the gpSEs are two *prognostic* equations for η and χ, we can numerically integrate them in this model to get their values at the next time-step if we wish, on the basis of a known complete state of the system at a particular time instant. But that would be an unnecessarily complicated way of evaluating the ageostrophic circulation in a 3D PE model since the following algorithm would be a far more straightforward way of computing the evolution of the complete flow:

(1) Time integration of (7.A4), (7.A5) and (7.A6) to get $(\hat{u} + \tilde{u})$, $(\hat{v} + \tilde{v})$ and $\hat{\theta}$ respectively at the next time-step.
(2) Use (7.A3a) and (7.A3b) to get $\hat{u}$ and $\hat{v}$ respectively at the next time-step.
(3) Evaluate $\tilde{u}$ and $\tilde{v}$ at the next time-step.
(4) Use (7.A7) to get $\tilde{w}$ at the next time-step.

Repeat.

8

Concluding Remarks

The illustrative analyses in the preceding chapters of this book convincingly demonstrate that *frontogenesis is an integral part of cyclogenesis*. A cyclone–front–convection system in the extratropics is embedded in a synoptic baroclinic wave consisting of a cold front and a warm front at the surface, together with an upper-level front near the tropopause. It is a location favorable for the development of severe weather elements. Figure 1.1 is a good example of such a system. A front is a hybrid in that its length scale is synoptic in the along-front direction and meso scale in the cross-front direction. Its synoptic scale stems from the intensifying baroclinic wave grown out of a background planetary flow. Its meso scale stems from a positive feedback mechanism critically depending on the direct advective influences of a cross-front ageostrophic circulation upon a geostrophic circulation of the intensifying baroclinic wave. The simplest mathematical representation of this process is the geostrophic momentum approximation. Therefore, atmospheric fronts are a product of *scale-interaction* involving planetary-, synoptic-, and meso-scale components of the flow, even in a dry model setting. In Chapter 4 we illustrated the formation of surface fronts from the instability of a highly idealized form of planetary-scale flow in mid-latitude (Eady flow or e-Charney flow) in the context of a 2D semi-geostrophic (SG) model. The synoptic scale of the emerging unstable baroclinic wave stems from the constraint of the earth's rapid rotation and strong background stratification. Fronts preferentially form at rigid boundaries where the normal velocity is constrained to zero. It is shown in Chapter 5 that a corresponding 3D SG model would simultaneously simulate cold and warm fronts.

Furthermore, we illustrate simultaneous formation of surface fronts and the upper-level front (ULF) at a dynamic tropopause during the intensification of an unstable baroclinic wave with the ARW-WRF model in Chapter 7. In the presence of moisture in this model, the meso-scale convective systems and small-scale convective elements associated with the fronts release the latent heat. They in turn significantly contribute to further intensifying the fronts. Thus, all scales of motion are active participants in the scale-interaction process through the wave–front–convection systems in this model, as in the actual atmosphere. The ULF gives rise to large local exchange of air of stratospheric and tropospheric origin. The ULF extends down to various tropospheric levels along it. The simulation reveals that the ULF and the surface cold front synergistically interact and can strengthen one another under favorable conditions. It has been found that such interaction would take place when the ULF is several hundred kilometers to the west of the surface cold front. We also verify that an appropriate measure for the development of a surface front is $F_2 \equiv \frac{D}{Dt}|\nabla_2\theta|$, whereas the rate of development of the ULF needs to be quantified in terms of $F_3 \equiv \frac{D}{Dt}|\nabla\theta|$. The details of the findings concerning the structure and dynamics of all fronts were summarized in Section 7.7. Those results are compatible with observations. These summary comments constitute a short answer to the question: "How do fronts form?"

Fronts may also be appreciated from the perspective of the global cascade of energy in the atmosphere. Under the influence of strong gravity, the compressibility of air would tend to give rise to a globally stable static stratification. The differential solar heating of the planet in the meridional direction on average additionally gives rise to a significant baroclinic zone in the mid-latitudes. There is a corresponding zonal baroclinic shear flow subject to the earth's strong rotation. Such a background baroclinic state of the atmosphere would be unstable with respect to synoptic-scale wave disturbances, which are in quasi-geostrophic (QG) balance. The development of such waves is an effective mode of energy transfer from the planetary scale to synoptic scale. The cascade of energy in a geostrophic turbulent flow, however, greatly diminishes toward the meso-scale range. That is an intrinsic characteristic of the geostrophic turbulence theory stemming from the simultaneous requirements of conservation of total energy and total enstrophy in such a flow. But this constraint would be broken by the atmospheric fronts in which the ageostrophic component of the flow plays a significant dynamic role. At the mature stage of fronts, when the local shear becomes strong and stratification becomes weak, meso-scale symmetric instability would further take place. A frontal system therefore provides an escape route for energy transfer from the synoptic- to meso-scale modes of motion in the atmosphere. In the presence of moisture, further cascade to the smaller scale would readily occur. Ultimately, the small-scale motions would be dissipated via friction and the related heat would be emitted back to space from the planet in the form of infrared radiation. That completes the energy cycle of the planet. In short, the development of fronts is an essential link in this overall scheme of energy cascade. It also has the effect of naturally maximizing the earth's entropy since an atmosphere having ubiquitous cyclone–front–convection systems is in a state of the highest possible degree of disorderliness. Fronts therefore play an important role in this fundamental issue as well. The discussion above from the general perspective of the Second Law of Thermodynamics might help address the question: "Why do fronts exist?"

It has taken many decades for atmospheric scientists to reach the current level of understanding of frontal dynamics. Over a century ago, Vilhelm Bjerknes used "Wrinkles in the Weather's Face" as a metaphor in his discussion of surface fronts. In the same spirit, it would be fitting to think of the ULFs metaphorically as "Veils in the Sky," which might locally shrink to closed curves at lower levels. Furthermore, we have a good idea as to when, how, and why a ULF (Veil) would synergistically interact with a surface cold front (Wrinkle).

Bibliography

Bergeron T. 1928: Uber die dreidimensional verknupfende Wettranalyse I. Geofys. *Publikasjoner*, 5, No. 6, 1–111.
1937: On the physics of fronts. *Bull. Amer. Meteor. Soc.*, 18, 265–275.
Bjerknes V. 1904: Das problem der wettervorhersge, betrachlet vom standpunkte der mechanic und der physic. *Meteorol. Zeitschr.*, 21, 1–7.
Bjerknes J, Solberg H. 1921: Meteorological conditions for the formation of rain. *Geofys. Publikasjner, Norske Videnskaps-Akad. Oslo*, 2, No. 3, 1–60.
Bjerknes V, Hesselberg Th, Devik O. 1911: *Dynamic Meteorology and Hydrography: Part II – Kinematics*. Carnegie Institution of Washington.
Bluestein HB. 1993: *Synoptic-Dynamic Meteorology in Midlatitudes*. Volume II. Oxford University Press.
Blumen W, Wu R. 1982: An analysis of Ekman boundary layer dynamics incorporating the geostrophic momentum approximation. *J. Atmos. Sci.*, 39, 1774–1782.
Bretherton FP. 1966: Critical layer instability in baroclinic flows. *Q. J. R. Meteorol. Soc.*, 92, 325–334.
Burger AP. 1962: On the non-existence of critical wave lengths in a continuous baroclinic stability problem. *J. Atmos. Sci.*, 19, 31–38.
Buzzi A, 2017: Comment on "Dynamics of Upper-Level Frontogenesis in Baroclinic Waves." *J. Atmos. Sci.*, 73, 309–312.
Buzzi A, Nanni A, Tagliazucca, M. 1977: Mid-tropospheric frontal zones: Numerical experiments with an isentropic coordinate primitive equation model. *Arch. Meteor. Geophys. Bioklim.*, A26, 155–178.
Buzzi A, Trevisan A, Salustri S. 1981: Internal frontogenesis: A two-dimensional model in isentropic semi-geostrophic coordinates. *Mon. Wea. Rev.*, 109, 1053–1060.
Charney JG. 1947: The dynamics of long waves in a baroclinic westerly current. *J. Meteor.*, 4, 135–163.
Charney JG, Eliassen A. 1964: On the growth of the hurricane depression. *J. Atmos. Sci.*, 21, 68–75.
Charney JG. 1975: Jacob Bjerknes : An appreciation. In *Selected Papers of Jacob Aall Bonneview Bjerknes*. M. Wurtele, ed. Western Periodical, 11–13.
Charney JG, Stern M. 1962: On the stability of internal baroclinic jets in a rotating atmosphere. *J. Atmos. Sci.*, 19, 159–172.
Chen, GTJ. 1996: The large-scale features associated with the Meiyu over South China and Taiwan. In *From Atmospheric Circulation to Global Change*. The Institute of Atmospheric Physics, Chinese Academy of Sciences, ed. China Meteorological Press, 289–313.
Cooper IM, Thorpe AJ, Bishop CH, 1992: The role of diffusive effects on potential vorticity in fronts. *Q.J.R. Meteorol. Soc.*, 118, 629–647.

Cullen MJP. 2006: *A Mathematical Theory of Large-Scale Atmosphere/Ocean Flow*. Imperial College Press.

Cullen MJP, Purser RJ. 1984: An extended Lagrangian theory of semi-geostrophic frontogenesis. *J. Atmos. Sci.*, 41, 1477–1497.

Danielson EF. 1968: Stratospheric-tropospheric exchange based on radioactivity, ozone and potential vorticity, *J. Atm. Sci.*, 25, 502–518.

Davies HC, Muller JC. 1988: Detailed description of deformation-induced semi-geostrophic frontogenesis. *Q. J. R. Meteorol. Soc.*, 114, 1201–1219.

Davies-Jones, R. 1991: The frontgenetical forcing of secondary circulation. Part I: The duality and generalization of the Q vector. *J. Atmos. Sci.*, 48, 497–509.

Doswell, CA III, Haugland MJ. 2007: A comparison of two cold fronts: Effects of the planetary boundary layer on the mesoscale. *Electron. J. Severe Storms Meteorol.* 2, 1–12.

Eady ET. 1949: Long waves and cyclone waves. *Tellus*, 1, 33–52

Eliassen A. 1948: The quasi-static equations of motion. *Geofys. Publ., Oslo*, 17, No. 3, 1–44.

1962: On the vertical circulation in frontal zones. *Geofys. Publ., Oslo*, 24, No. 4, 147–160.

Emanuel KA. 1994: *Atmospheric Convection*. Oxford University Press.

Emanuel KA, Fantini M, Thorpe AJ. 1987: Baroclinic instability in an environment of small stability to slantwise moist convection. Part I: Two-dimensional models. *J. Atmos. Sci.*, 44, 1559–1573.

Faller AJ. 1956: A demonstration of fronts and frontal waves in atmospheric models. *J. Meteor.*, 13, 1–4.

Fedorov KN. 1984: *Characteristics Features of Oceanic Fronts*. Springer-Verlag.

Fultz D. 1952: On the possibility of experimental models of the polar-front wave. *J. Meteor.* 9, 379–384.

Gill AE. 1982: *Atmosphere–Ocean Dynamics*. Academic Press.

Green JSA, 1960: A problem in baroclinic instability. *Q. J. R. Meteorol. Soc.*, 86, 237–251.

Green J. 2004: *Atmospheric Dynamics*. Cambridge University Press.

Hakim GJ, Keyser D. 2001: Canonical frontal circulation patterns in terms of Green's functions for the Sawyer–Eliassen equation. *Q. J. R. Meteorol. Soc.*, 127, 1795–1814.

Haltiner GJ, Martin FL. 1957: *Dynamical and Physical Meteorology*. McGraw-Hill.

Heckley WA, Hoskins BJ. 1982: Baroclinic waves and frontogenesis in a non-uniform potential vorticity semi-geostrophic model. *J. Atmos. Sci.*, 39, 1999–2016.

Hines KM, Mechoso CR. 1991: Frontogenesis processes in the middle and upper troposphere. *Mon. Wea. Rev.*, 119, 1225–1241.

Holton J. 1992: *Introduction to Dynamic Meteorology*. Academic Press.

Hong SY, Noh Y, Dudhia J. 2006: A new vertical diffusion package with an explicit treatment of entrainment processes. *Mon. Wea. Rev.*, 134, 2318–2341.

Hoskins BJ. 1972: Non-Boussinesq effects and further development in a model of upper level frontogenesis. *Quart. J. R. Met. Soc.*, 98, 532–541.

1975: The geostrophic momentum approximation and the semi-geostrophic equations. *J. Atmos. Sci.*, 32, 233–242.

1976: Baroclinic waves and frontogenesis. Part I. Introduction and Eady waves. *Q. J. R. Meteorol. Soc.*, 102, 103–122.

Hoskins BJ, Bretherton FP. 1972: Atmospheric frontogenesis models: Mathematical formulation and solutions. *J. Atmos. Sci.*, 29, 11–37.

Hoskins BJ, Draghici I. 1977: The forcing of ageostrophic motion according to the semi-geostrophic equations and in an isentropic coordinate model. *J. Atmos. Sci.*, 34, 1859–1867.

Hoskins BJ, Heckley WA. 1981: Cold and warm fronts in baroclinic waves. *Q. J. R. Meteorol. Soc.*, 107, 79–90.

Hoskins BJ, James IN. 2014: *Fluid Dynamics of the Midlatitude Atmosphere*. Wiley – Blackwell.

Hoskins BJ, West NV. 1979: Baroclinic waves and frontogenesis. Part II: Uniform potential vorticity jet flows - cold and warm fronts. *J. Atmos. Sci.*, 36, 1663–1680.

Jewell R. 1994: The integrity at work in the Bergen School of meteorology. In *The Life Cycles of Extratropical Cyclones*. Volume I, Gronas, S., Shapiro MA, ed. American Meteorological Society, 32–37.

Keyser D, Pecnick M. 1985a: A two-dimensional equation model of frontogenesis forced by confluence and horizontal shear. *J. Atmos. Sci.*, 42, 1259–1282.

1985b: Diagnosis of ageostrophic circulations in a two-dimensional primitive equation model of frontogenesis. *J. Atmos. Sci.*, 42, 1283–1305.

1987: The effect of along-front temperature variation in a two-dimensional primitive equation model of surface frontogenesis. *J. Atmos. Sci.*, 44, 577–604.

Keyser D, Shapiro MA. 1986: A review of the structure and dynamics of upper-level frontal zones. *Mon. Wea. Rev.*, 114, 452–499.

Keyser D, Schimdt D, Duffy G. 1989: A technique for representing three-dimensional vertical circulations in baroclinic disturbances. *Mon Wea. Rev.*, 117, 2463–2494.

Kuo, HL. 1973: Dynamics of quasi-geostrophic flows and instability theory. *Adv. Appl. Mech.*, 13, 247–330.

Lackmann G. 2011: *Midlatitude Synoptic Meteorology: Dynamics, Analysis & Forecasting*. American Meteorology Society.

Lapeyre G, Held IM. 2004: The role of moisture in the dynamics and energetics of turbulent baroclinic eddies. *J. Atmos. Sci.*, 61, 1693–1710.

Lin YL. 2008: *Mesoscale Dynamics*. Cambridge University Press.

Lindzen RS. 1990: *Dynamics in Atmospheric Physics*. Cambridge University Press.

Magnusdottir G, Schubert WH. 1990: On the generalization of semigeostrophic theory to the β-plane. *J. Atmos. Sci.*, 47, 1714–1720.

1991: Semigeostrophic theory on the hemisphere. *J. Atmos. Sci.*, 48, 1449–1456.

Mak M. 1994: Cyclogenesis in a conditionally unstable moist baroclinic atmosphere. *Tellus*, 46A, 14–33.

2011: *Atmospheric Dynamics*. Cambridge University Press.

2014: Life cycle of fronts: Intrinsic dynamics in a 3D semi-geostrophic model. *Q. J. R. Meteorol. Soc.*, 140, 2211–2222.

Mak M, Bannon PB. 1984: Frontogenesis in a moist semi-geostrophic model. *J. Atmos. Sci.*, 41, 3485–3500.

Mak M, Lu Y, Deng Y. 2016: Dynamics of upper-level frontogenesis in baroclinic waves. *J. Atmos. Sci.*, 73, 2699–2714.

2017: Comments on "dynamics of upper-level frontogenesis in baroclinic waves." *J. Atmos. Sci.*, 74, 313–315.

2018: Two issues concerning surface frontogenesis. *J. Atmos. Sci.*, (in press).

Markowski, P, Richardson Y. 2010: *Mesoscale Meteorology in Midlatitudes*. Wiley-Blackwell.

Marshall J, Plumb RA. 2007: *Atmosphere, Ocean and Climate Dynamics*. Elsevier Academic Press.

Martin JE. 2006: *Mid-latitude Atmospheric Dynamics: A First Course*. Wiley and Sons.

Martin JE, Locatelli JD, Hobbs PV. 1993: Organization and structure of clouds and precipitation on the Mid-Atlantic Coast of the United States. Part I: The synoptic evolution of a deep tropospheric frontal circulation and attendant cyclogenesis. *Month. Wea. Rev.* 121, 1299–1316.

McWilliams JC. 2006: *Fundamentals of Geophysical Fluid Dynamics*. Cambridge University Press.

Montgomery MT, Farrell BF. 1990: Dry surface frontogenesis arising from interior potential vorticity perturbations in a semigeostrophic model. *J. Atmos. Sci.*, 47, 2837–2852

1991: Moist surface frontogenesis associated with interior potential vorticity anomalies in a semigeostrophic model. *J. Atmos. Sci.*, 48, 343–367.

Morgan MC. 1989: Optimal excitation of baroclinic waves. *J. Atmos. Sci.*, 46, 163–172.
Mudrick SE. 1974: A numerical study of frontogenesis. *J. Atmos. Sci.*, 31, 869–892.
Nakamura N. 1994: Nonlinear equilibration of two-dimensional Eady waves: simulations with viscous geostrophic momentum equations. *J. Atmos. Sci.*, 51, 1023–1035.
Nakamura N, Held IM. 1989: Nonlinear equilibration of two-dimensional Eady waves. *J. Atmos. Sci.*, 46, 3055–3064.
Newton CW. 1954: Frontogenesis and frontolysis as a three-dimensional process. *J. Meteor.* 11, 449–461.
1958: Variations in frontal structure of upper level troughs. *Geophysica*, 6, 357–375.
Newton CW, Trevisan A. 1984: Clinogenesis and frontogenesis in set-stream waves. Part II: Channel model numerical experiments. *J. Atmos. Sci.*, 41, 2735–2755.
Nieman PJ, Martin Ralph D, Shapiro MA. 1998: An observational study of fronts and frontal mergers over the continental United States. *Mon. Wea. Rev.*, 126, 2521–2553.
Ninomiya, K. 1984: Characteristics of Baiu front as a predominant subtropical front in the summer northern hemisphere. *J. Meteor. Soc. Japan*, 62, 880–893.
O'Gorman PA. 2011: The effective static stability experienced by eddies in a moist atmosphere. *J. Atmos. Sci.*, 68, 75–90.
Orlanski I. 1968: Instability of frontal zones. *J. Atmos. Sci.*, 25, 178–200.
Orlanski I, Ross BB. 1977: The circulation associated with a cold front. Part I: Dry case. *J. Atmos. Sci.*, 34, 1619–1633.
Palmen E, Newton CW. 1969: *Atmospheric Circulation Systems*. Academic Press.
Pedlosky J. 1987: *Geophysical Fluid Dynamics*, 2nd edn. Springer-Verlag.
Pettersen S. 1956: *Weather Analysis and Forecasting*. Volume I, 2nd edn. McGraw-Hill.
Phillips NA. 1956: The general circulation of the atmosphere: A numerical experiment. *Quart. J. Roy. Meteor. Soc.*, 82, 123–164.
Plougonven R, Snyder C. 2007: Inertia-gravity waves spontaneously generated by jets and fronts. Part I: Different baroclinic life cycles. *J. Atmos. Sci.*, 64, 2502–2520.
Reed RJ. 1955: A study of a characteristic type of upper-level frontogenesis. *J. Meteor.* 12, 226–237.
Reed RJ, Danielson EF. 1959: Fronts in the vicinity of the tropopause. *Arch. Meteor. Geophys. Bioklim.*, A11, 1–17.
Reed RJ, Sander F. 1953: An investigation of the development of a mid-tropospheric frontal zone and its associated vorticity field. *J. Meteor.*, 10, 338–349.
Richardson LF. 1922: *Weather Prediction by Numerical Process*. Cambridge University Press, reprinted by Dover in 1965.
Rotunno R, Bao JW. 1996: A case study of cyclogenesis using a model hierarchy. *Mon. Wea. Rev.*, 124, 1051–1066.
Rotunno R, Skamarock WC, Snyder C. 1994: An analysis of frontogenesis in numerical simulations of baroclinic waves. *J. Atmos. Sci.*, 23, 3373–3398.
Salmon R. 1985: New equations for nearly geostrophic flow. *J. Fluid Mech.*, 153, 461–477.
1998: *Lectures on Geophysical Fluid Dynamics*. Oxford University Press.
Sanders F. 1955: An investigation of the structure and dynamics of an intense surface frontal zone. *J. Meteor.*, 12, 542–552.
Sawyer JS. 1956: The vertical circulation at meteorological fronts and its relation to frontogenesis. *Proc. R. Soc. London, Ser. A.*, 234, 346–362.
Schar C, Wernli H. 1993: Structure and evolution of an isolated semi-geostrophic cyclone. *Quart. J. Roy. Meteorol.Soc.*, 119, 57–90.
Schubert WH. 1985: A note on semi-geostrophic theory. *J. Atmos. Sci.*, 42, 1770–1772.
Shapiro MA. 1975: Simulation of upper level frontogenesis with a 20-level isentropic coordinate primitive equation model. *Mon. Wea. Rev.*, 103, 591–604.
1978: Further evidence of the mesoscale and turbulent structure of upper level jet stream-frontal zone system. *Mon. Wea. Rev.*, 106, 1100–1111.

1981: Frontogenesis and geostrophically forced secondary circulations in the vicinity of jet stream-frontal zone system. *J. Atmos. Sci.*, 38, 954–973.

1982: *Mesoscale Weather Systems of the Central United States*. CIRES/NOAA Technical Report. University of Colorado.

Shapiro MA, Keyser D. 1990: Fronts, jet streams, and the tropopause. *Amer. Meteor. Soc.*, Extratropical Cyclones: The Erik Palmen Memorial Volume, 167–191.

Shapiro MA, Hample T, van de Kamp DW. 1984: Radar wind profiler observations fronts and jet streams. *Mon. Wea. Rev.* 112, 1263–1266.

Shutts GJ. 1989: Planetary semi-geostrophic equations derived from Hamilton's principle. *J. Fluid Mech.*, 208, 545–573.

Sinclair VA, Keyser D. 2015: Force balances and dynamical regimes of numerically simulated cold fronts within the boundary layer. *Q. J. R. Meteorol. Soc.*, 141, 2148–2164.

Skamarock WC, Klemp J, Dudhia J, et al. 2008: *A Description of the Advanced Research WRF Version 1*. NCAR Technical Note 4751STR. National Center for Atmospheric Research.

Snyder C, Skamarock WC, Rotunno R. 1991: A comparison of primitive-equation and semigeostrophic simulations of baroclinic waves. *J. Atmos. Sci.*, 48, 2179–2194.

1993: Frontal dynamics near and following frontal collapse. *J. Atmos. Sci.*, 50, 3194–3211.

Tan Z, Zhang F, Rotunno R, Snyder C. 2004: Mesoscale predictability of moist baroclinic waves: Experiments with parameterized convection. *J. Atmos. Sci.*, 61, 1794–1804.

Thorpe AJ, Nash CA. 1984: Convective and boundary layer parameterizations in a diagnostic model of atmospheric fronts. *Q. J. R. Meteorol. Soc.*, 110, 443–466.

Trapp RJ. 2013: *Mesoscale-Convective Processes in the Atmosphere*. Cambridge University Press.

Trapp RJ, Diffenbaugh NS, Brooks HW. 2007: Changes in severe thunderstorm environment frequency during the 21st century caused by anthropogenically enhanced global radiative forcing. *Proc. Natl. Acad. Sci.*, 104, 19719–19723, doi:10.1073/pnas.0705494104.

Trenberth KE. 1998: Atmospheric moisture residence times and cycling: Implications for rainfall rates and climate change. *Climatic Change*, 39, 667–694.

Uccellini, LW, Johnson DR. 1979: The coupling of upper and lower tropospheric jet streaks and implications for the development of severe convective storms. *Mon. Wea. Rev.*, 107, 682–703.

Uccellini LW, Keyser D, Brill KF, Wash CH. 1985: The Presidents' Day cyclone of 18–19 February 1979: Influence of upstream trough amplification and associated tropopause folding on rapid cyclogenesis. *Mon. Wea. Rev.*, 113, 962–988.

Vallis GK. 2006: *Atmospheric and Oceanic Fluid Dynamics: Fundamentals and Large Scale Circulation*. Cambridge University Press.

von Arx WS, Bumpus DF, Richardson WS. 1955: On the fine structure of the Gulf Stream. *Deep-Sea Res.* 3, 46–65.

Wei J, Zhang F. 2014: Mesoscale gravity waves in moist baroclinic jet-front system. *J. Atmos. Sci.*, 71, 929–952.

Williams RT. 1967: Atmospheric frontogenesis: A numerical experiment. *J. Atmos. Sci.*, 24, 627–641.

1968: A note on quasi-geostrophic frontogenesis. *J. Atmos. Sci.*, 25, 1157–1159.

Williams RT, Plotkin J. 1968: Quasi-geostrophic frontogenesis. *J. Atmos. Sci.*, 25, 201–206.

Yudin MI. 1955: Invariant quantities in large-scale atmospheric processes. *Tr. Glav. Geofiz. Observ.*, 55, 3–12.

Zhang F. 2004: Generation of mesoscale gravity waves in the upper-tropospheric jet-front systems. *J. Atmos. Sci.*, 61, 440–457.

Index

Printed by Integrated Books International,
United States of America